VENOM

VENOM

The Secrets of Nature's Deadliest Weapon

Ronald Jenner & Eivind Undheim

Smithsonian Books
Washington, DC

For Sarah and Hanne

for cheerily (mostly) enduring the relentless barrage of
venom 'facts of the day' and for being there, always

Published in Great Britain by the Natural History Museum,
Cromwell Rd, London SW7 5BD

Copy-edited by Celia Coyne
Designed by Mercer Design, London
Reproduction by Saxon Digital Services

Published in North America by Smithsonian Books

This book may be purchased for educational, business, or sales
promotional use. For information, please write: Special Markets
Department, Smithsonian Books, P.O. Box 37012, MRC 513,
Washington, DC 20013

ISBN 978-1-58834-454-0

Printed in China by Toppan Leefung Printing Limited, not at
government expense

21 20 19 18 17   5 4 3 2 1

Contents

Chapter 1

Nature's Ultimate Weapon

It is hard to escape them. You may not see them, but *they* are there, no matter where *you* are. You may be camping in a forest or picnicking in a field. You may be snorkelling in the sea or just strolling through the city. You may not notice them, but they are near. The old adage that you're never very far from a rat may not strictly be true (lab technicians and sewer flushers excepted). But certainly you are unlikely to ever be out of sight of a venomous animal.

'I'm so glad that I don't live in the Amazon!' you may think at this point. Alas, your proximity to venomous creatures is virtually guaranteed in practically every inhabitable place on Earth, unless you've retreated to an Antarctic research station far away from the ocean's edge. Even if you are just pottering about in your kitchen or garden, venomous animals will be within only a few metres of you. You might be in London or Brazzaville, São Paulo or Tyrol, it makes no difference.

When asked to picture a venomous animal, the average reader is likely to conjure up an image of a snake curled on a branch, a spider hanging in its web, or a scorpion poised to strike. Such imagery triggers strong emotions. Venomous animals are icons of mortal danger. We rightly regard them with a mixture of awe, fear and respect, but our curiosity drives us closer to have a better look. We know

A black mamba signals potential enemies to beware by opening its mouth and showing its dark lining.

After injecting a paralyzing venom into the goldfish with its proboscis, this giant water bug, *Lethocerus* sp., sucks up its liquified meal.

that their stillness is a ruse. When unsuspecting prey comes close enough, the predators break from their frozen posture with explosive speed, striking out with venomous fangs or stingers to envenomate their prey.

Once stung or bitten, it is typically game over. The toxin cocktails that make up predatory venoms are extraordinarily powerful weapons that have evolved to subdue prey before it has a chance to escape or mount a defensive attack. But venom isn't just used for catching prey. It is the defensive use of venom that lies at the heart of our own fear for these toxic assailants, and with good reason. Some snakes, spiders and scorpions have venoms that are potent enough to kill animals orders of magnitude larger than themselves. Snakes envenomate an estimated 1.8 million people each year, resulting in about 400,000 amputations and 100,000 deaths. Scorpions and spiders kill another 5,000 or so people every year. The number of deaths directly accountable to envenomations by other venomous animals, such as bees, wasps and jellyfish, are smaller, but venomous animals also cause human deaths indirectly through the transmission of diseases. Venomous blood-feeding triatome 'kissing' bugs and mosquitoes, for instance, are disease vectors that cause untold suffering and hundreds of thousands of deaths each year. But shocking as these figures may be, the venomous animals responsible for causing human suffering represent only one small facet of the extraordinary diversity of the world of venom.

Venom as a force for good

Venomous animals play important ecological roles in their local habitats, and they provide many benefits that are woven into our own lives. Venomous predators help regulate the abundance and ecological impact of prey species. Without snakes and spiders and other venomous predators, we might be overrun by roiling waves of rodents and suffocate in clouds of insects, not least in areas of human habitation where our food and waste products attract a great many unwanted guests.

Venomous animals also help safeguard our food sources, for instance by limiting the destruction that herbivorous insects can inflict upon crops. Parasitoid wasps use their sophisticated venoms to turn countless leaf-munching caterpillars, sap-sucking bugs and voracious beetle grubs into living meat larders for their larvae. The venoms act on the hosts in various ways: inhibiting their immune systems, arresting their development, causing paralysis, dissolving their tissues and changing their physiology so that they make more nutrients available to be consumed by the wasp larvae, and farmers and gardeners gladly harness this parasitoid power.

Venomous animals even provide important food sources themselves. The next time you are enjoying a dish of Southern calamari in a Sydney restaurant you may want to reflect on the fact that you are eating a venomous predator. *Sepioteuthis australis*, as this species is known scientifically, is just one of about 290 species of squid, which together with their cuttlefish and octopus cousins, contribute to the 1–4 million tons of cephalopods caught annually worldwide. The venom of the Southern calamari is a toxic cocktail, one component of which is a neurotoxin that causes paralysis and death in crabs, a favourite cephalopod prey. Gram for gram this squid venom toxin is as deadly to crabs as the most lethal snake venom toxins are to mice. With the exception of the handful of living *Nautilus* species, all cephalopods are thought to be venomous. This makes a monstrous mollusc – the colossal squid *Mesonychoteuthis hamiltoni* – the prime contender for being the largest and heaviest venomous animal on Earth.

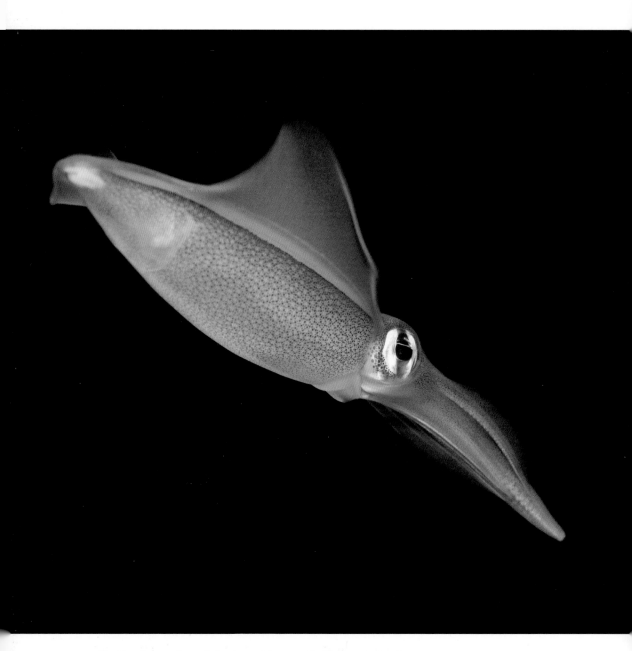

The Southern calamari, *Sepioteuthis australis,* is a streamlined venomous predator that is adept at catching fast fish and crustaceans.

As this book will show, the value of venomous animals is not limited to assisting in securing global food sustainability. The enormous diversity of venomous animals – with well over 200,000 known venomous species – can help us face many challenges of the modern world, on a local and global scale, but their value is easily overlooked. On the local end of the spectrum, worm diggers working along Maine's eastern seaboard in the USA are butting heads with clam diggers to assert their legal rights to dig for bloodworms, a group of venomous polychaetes that sustains a local, multimillion-dollar fishing-bait industry. The clam diggers would prefer the worm diggers to be barred from the muddy intertidal flats to allow juvenile clams to seed and grow undisturbed.

Unbeknownst to most worm diggers, who do occasionally suffer a painful bite that can swell a hand like a balloon, bloodworm venom has for years provided neuroscientists with a tool to investigate how nerves communicate with muscles. The venom of a European species of bloodworm, *Glycera tridactyla*, which is likewise a popular bait for sea fishing, contains a unique neurotoxin called glycerotoxin. This toxin causes a massive and long-lasting release of the neurotransmitter acetylcholine into the junction between nerve and muscle cells, which results in uncontrollable muscle twitching. Because this effect is reversible, neuroscientists are using this toxin to study how nerves function.

On the global end of the spectrum, venom is at the heart of the best-selling anti-diabetes drugs Byetta and Bydureon. In 2015 the pharmaceutical company AstraZeneca reported US$896 million of combined worldwide sales of the drugs. The profits made on the sales of such blockbuster anti-diabetes drugs are steadily increasing in the wake of the type II diabetes epidemic that trails the westernization of the world's diet. Byetta and Bydureon are synthetic versions of a peptide, which is a small protein, found in the venom of the Gila monster, *Heloderma suspectum*, found in the southwest United States and northwest Mexico. Known as exendin-4, this peptide potently lowers glucose levels in the blood, which may explain its role in the life of this gluttonous lizard.

Gila monsters are binge-eating champions. They can survive on large, infrequent meals, and they can store large amounts of fat in their tails to survive periods of food scarcity. When chomping down on a large meal, or even when forced to bite on a stick for a few minutes, they increase the level of exendin-4 in their blood, helping them to process the abundance of absorbed nutrients. When injected into a human patient, the synthetic analogue of the lizard peptide performs the same role. So by mimicking a venom component that enables a lizard to eat to excess when the opportunity arises, Byetta and Bydureon help us to deal with the consequences of our own insatiable appetites.

The venom of the Gila monster, *Heloderma suspectum*, was the unlikely source for the development of two powerful drugs for the treatment of type 2 diabetes.

Venom, poison or toxin?

In 2014 one of the authors was involved in the publication of a scientific paper that proved for the first time the existence of venomous crustaceans. The research showed that the centipede-like aquatic crustaceans known as remipedes are able to produce and inject venom to catch their elusive prey in pitch-dark aquatic caves. One website reported the discovery in this way: 'World's first poisonous crab found in Western Australia!' Quite apart from the fact that the blogger failed to realize that the species investigated was neither a crab, nor was it from Australia, it illustrates the common confusion between what is venomous and what is poisonous.

A toxin is any toxic substance, irrespective of whether it is a poison or part of a venom (venoms are generally cocktails of toxins). Venom is broadly defined as a toxic secretion produced by specialized cells in one animal that is delivered to another animal via a delivery mechanism – typically through the infliction of a wound – to disrupt normal physiological functioning in the interest of predation, feeding, defence, competition or other biological processes that benefit the venom-producing animal. Venom delivery may be active, such as a snake biting a mouse or a bee stinging you; or it may be passive, as in the case of a swimmer stepping on a stonefish. Predatory venoms are always delivered actively, but many defensive envenomations are passive.

Poison, on the other hand, is a toxic substance that is passively transferred without the involvement of a delivery mechanism or infliction of a wound, usually through ingestion, inhalation or absorption through the skin.

Given that the delivery of toxins via a wound, however tiny, is a hallmark of venoms, blood-feeding animals such as leeches, mosquitoes, ticks, vampire bats, triatome 'kissing' bugs and even vampire snails should also be considered venomous. Their venoms have evolved to ensure a steady supply of flowing blood and to minimize the chances of being detected by their hosts when they are feeding.

Although the delivery mode is the primary definition criterion, there are three subsidiary criteria to help distinguish poison and venom. First, venoms and poisons

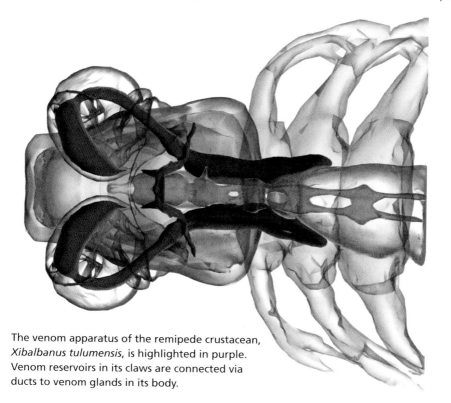

The venom apparatus of the remipede crustacean, *Xibalbanus tulumensis*, is highlighted in purple. Venom reservoirs in its claws are connected via ducts to venom glands in its body.

have different 'signature' components. Typically the most abundant and functionally important venom ingredients are peptides and proteins. Because stomach acid denatures and inactivates proteins and peptides it is generally harmless to swallow venom, as long as the venom cannot penetrate the body through wounds in the mouth or oesophagus. In contrast, poisons are typically organic compounds other than peptides and proteins, such as alkaloids (e.g. nicotine, caffeine or tetrodotoxin), that occur singly or in a mixture. Second, whereas venoms have evolved to serve a host of functions, poisons are almost exclusively defensive. Third, venoms are almost exclusively produced endogenously, by the animal that is using the venom itself. In contrast, poisons are typically of exogenous origin: they are synthesized by one organism, and then harvested, often by eating them, by another organism. Many poisons are ultimately derived from plants or microorganisms, even though they may serve to defend the lives of insects, amphibians or other animals.

Poison dart frogs, *Dendrobates* spp., from Central and South America, and mantella frogs, *Mantella* spp., from Madagascar, are classic examples of poisonous animals. All are small and their bright colours serve as warning signals to potential predators. The brighter and more exuberantly coloured poison dart frogs are also more toxic. These frogs sequester powerful chemicals in their skin, which they obtain by eating arthropods packed with alkaloid toxins, such as mites, beetles, millipedes and ants. Of course the drawback of relying on defensive toxins derived from food is that these animals will detoxify when they eat a harmless diet.

Definitions, by definition, are exclusive, and Mother Nature often resists being parsed into neat categories. Consider the alkaloid tetrodotoxin (TTX). Eating pufferfish sushi is risky because the fish sequesters this potent neurotoxin, which is of bacterial origin, in its internal organs, especially the ovaries and liver. Eating pufferfish tainted with TTX can cause respiratory paralysis and death since TTX

The toxic alkaloids sequestered by the green poison dart frog, *Dendrobates auratus*, are an effective defence against predators, such as the large tarantula, *Sericopelma rubronitens*.

blocks the transmission of nerve impulses along nerve fibres, and muscles, including the diaphragm, cannot contract if they fail to receive nerve impulses. As a non-proteinaceous, exogenously produced, passively transferred, defensive compound, TTX is therefore a textbook example of a poison.

But what about the blue-ringed octopus, *Hapalochlaena* spp.? These animals likewise contain lethal doses of TTX. If a 110 kg (243 lb) herbivorous green sea turtle accidentally ingests a tiny 4 cm (1½ in) long blue-ringed octopus while grazing on sea grass, it will become paralyzed and drown. However, the octopus can also actively deliver its TTX via a bite that can kill a human. TTX is therefore both a poison and a venom depending on its method of delivery. Being able to sequester TTX is clearly of great adaptive value. It is present in a wide range of animals in several phyla, including some newts, echinoderms (sea stars and relatives), ribbon worms (Nemertea), flatworms (Platyhelminthes) and arrow worms (Chaetognatha). Interestingly, some of the species in the latter two phyla use TTX as a predatory toxin. For example, one undescribed tropical species of marine flatworm has a high

A blue-ringed octopus will flash its blue rings as a warning signal when disturbed or harassed. If that doesn't deter the attacker, it may deliver its lethal bite.

A cluster of the geophilomorph centipede *Strigamia maritima* from the Scottish coast. In contrast to most other centipedes this species is gregarious, which makes them easy to collect for scientific research.

concentration of TTX in its pharynx. The flatworm feeds on snails, which are quickly killed when they come into contact with TTX.

Some animals that are both poisonous and venomous use different substances for each. The Asian tiger keelback snake, *Rhabdophis tigrinus,* is a predator that uses its venom to subdue its prey, which are typically frogs and toads. Although its venom is strong enough to kill a human, by interfering with blood clotting and causing internal bleeding, the snake's typical defence is to release highly cardiotoxic steroids from glands in its neck. The snakes obtain this poison from eating toxic toads. Remarkably, pregnant females change their foraging behaviour to seek out toads so they can provision their eggs with the toxins, providing the baby snakes with a chemical defence from the moment of hatching.

Geophilomorph centipedes (soil centipedes) are another example of simultaneously poisonous and venomous animals. Although these arthropods have venom claws on their head to dispatch prey with, many also have noxious glands on their belly surface from which they can secrete a sticky cyanogenic liquid. When expelled, this liquid generates hydrogen cyanide gas, a universally deadly poison that halts cellular respiration. Clearly it is advantageous for a long-bodied soil dweller to have a defensive strategy that is not restricted to its head end, and that is able to deter predators such as ants and spiders.

Dissecting Nature's ultimate weapon

When we enter the world of venom, we enter the realm of one of the most diverse, versatile, sophisticated and deadly biological adaptations ever to have evolved on the planet. Venom truly is Nature's ultimate weapon, the result of an evolutionary arms race that escalated into chemical warfare. The astonishing adaptive value of venom is demonstrated by the fact that venom has evolved almost 90 times in the animal kingdom. These venomous ancestors have given rise to the more than 200,000 venomous animal species that are known today. In fact, venom systems – venoms and the associated anatomical and behavioural traits needed for venom delivery – are the oldest animal offensive weapons to have evolved on Earth. This early evolutionary step is represented by the stinging cells of cnidarians. With a fossil record going back at least 600 million years, cnidarians are the oldest lineage of venomous animals. Indeed, they received their name for their cnidae, the complex cell organelles that deliver their venom explosively to victims.

Males of the skeleton shrimp, *Caprella scaura,* often engage in lethal battles over females, in which they try to envenomate each other with a venom tooth located on each of their large claws. The venom tooth (highlighted) is the one closest to the articulation of the claw's blade-like terminal segment.

One important factor that helps explain the huge variety of venomous animals and the rampant convergent evolution of venoms is that venom cocktails contain elements produced from the genomic toolkit that is shared by all animals. This is strikingly revealed by the many instances in which venomous animals have independently recruited the same genes to produce venom toxins. The genome of any animal, whether it is a slow loris or the tiny skeleton shrimp, has the genes necessary to evolve venom toxins, given the appropriate selective conditions. And indeed, both the slow loris and skeleton shrimp are venomous.

Of course ammunition without a gun doesn't make a weapon, and having toxins doesn't make an animal venomous by itself. Remarkably, it appears that almost any conceivable animal body plan has the flexibility to evolve the equipment to effectively and efficiently deliver venom, be they the bones of a frog skull, the radula of a mollusc, the tail of a scorpion or the spur of a platypus. Clearly, venom systems, both gun and ammunition, have been fashioned from ubiquitous building blocks.

However, venom systems aren't stocked with identical toxins. A venom acts more like a battalion of snipers than as a machinegun loaded with one type of bullet. Evolution has honed venom toxins into extremely target-specific compounds, each of which disrupts the physiological functioning of the victim in specific ways. Because the most complex venoms contain hundreds or even thousands of distinct components, they are able to overcome the defences of almost any victim, and venoms are often effective in both defence and predation. And because venom glands can replenish their contents, venom systems have a near limitless magazine of ammunition.

Moreover, venom toxins act as self-guided bullets. Even when venom is delivered to a sub-optimal location in the body of a prey or predator, the toxins will find their targets. Each toxin generally only has a high affinity for a specific target, and until the toxin encounters this target, it will course through the body unhindered.

Yet, chemically complex venom doesn't require a cumbersome reservoir. Venom is used with equal facility by the colossal squid, probably the largest invertebrate in the world, and the tiniest parasitoid wasp. Baby venomous animals are often born or hatch with a fully functional venom system. With venom as your weapon, you don't need to grow large to be able to overpower your prey or foe by sheer strength, or to learn the complexities of a lethal technique that relies on close proximity combat. The wet weight of snake, spider or scorpion venom is generally less than 0.5% of the venom-producing animal's body weight. Translated into human terms, an average man of about 70 kg (154 lb) has about 28 kg (61 lb) of skeletal muscle (about 40% of total body mass) that can be fashioned into a lethal weapon. However, if that man was venomous, he would be able to achieve the same lethal potential by producing just 350 g (12 oz) of venom. However, this kind of power doesn't come cheap. For some venomous snakes the metabolic costs of replenishing depleted venom glands equal those of shedding their skin.

The spitting spider, *Scytodes thoracica,* ejects a mixture of glue and venom from its venom fangs with a speed of 100 km per hour (62 mph) while vibrating its fangs 1,700 times per second. This traps prey under a neat and sticky zig-zag mesh in significantly less time than it takes to blink.

But there are even more qualities that make venom the ultimate biological weapon. The act of envenomation generally happens very rapidly, and often requires minimal contact with the target, thereby minimizing the risk of injury to the venomous animal. The rapid strike of a snake, or the sting of a cone snail can overwhelm prey in a matter of seconds, while the spitting of venomous silk by a spider or the spraying of toxins by ants can incapacitate prey or predator without requiring contact at all.

Finally, venom systems have evolved to work in all environments. They silently deliver their toxic ammunition in the deepest oceans and the driest deserts, high in the air and deep underground. Venoms topple elephants in Africa and fell flies in your house. They allow snails to eat fish and centipedes to dine on bats. They allow leeches to suck blood and the social insects to dominate terrestrial ecosystems. They are a key factor in the evolutionary success of countless species all over the world. But they also cause indescribable human suffering. Snakes envenomed approximately 35 people worldwide in the time it took you to read this far into the book. But through their proven value and future promise in the development of new pharmaceuticals, venoms also help us face the many challenges of modern life, from combating disease to safeguarding food crops. Venom is truly Nature's ultimate weapon.

Structure of the book

The chapters that follow explore the many hidden facets of the world of venom. Chapter 2 introduces the extraordinary diversity of venoms. We visit the furthest corners of the world of venom to illustrate the extreme diversity of animals that use venoms and examine the many ways that venoms are delivered. We then explore the surprising range of roles that venoms play in the lives of venomous animals, including predation, defence, mating, communication, competition, reproduction and even home building.

Chapter 3 describes how venoms are studied by scientists, followed by a detailed look at how the power of venom resides in the complexity of toxin cocktails in chapter 4. Venoms can contain hundreds or even thousands of unique peptides and proteins, which home in on a large number of targets in the body of the victim. As a result, venoms are able to interfere dramatically with the normal physiological functioning of the envenomated victim. Some of the toxic effects of venom are

The jararaca pit viper, *Bothrops jararaca*, embodies the good and the bad of venom. Although it is responsible for many snakebites in South America, its venom has been used to develop a potent drug for the treatment of high blood pressure.

subtle and enormously sophisticated, while others are brute force attacks that destroy the functional integrity of the victim irreversibly. These chapters show what makes venoms toxic to the blood, nervous system, heart, muscles and cells and tissues, and how these toxicities confer great adaptive value on venoms.

Chapter 5 looks at how venoms evolve. Although scientists can determine the toxin composition of venom with great precision, a venom cocktail is not a static concoction. Closely related species may have completely different venom cocktails, and the make-up of venoms can even change substantially during the lifetime of a single individual. On the other hand, striking similarities in venom composition have evolved in distantly related groups, revealing some deep generalities of venom evolution as venom evolves in the context of the ongoing evolutionary arms race between predator and prey.

Chapter 6 shows how venom has long intersected with aspects of human culture and our daily lives, such as the veneration of venomous animals by the ancients, the use of venomous animals in rituals and traditional medicine, and the development of new life-changing pharmaceuticals. It also focuses attention on the human suffering caused by venomous animals, especially as a result of snakebite, as the good and the bad of venom are inextricably linked. And finally, the world of venom is encapsulated in a nutshell through the microcosm of honeybees in chapter 7. Our snake and spider phobias are a hint that our lineage has long evolved in the company of these venomous and sometimes dangerous creatures. But as we come to better understand the beautiful intricacies of the biology and evolution of venoms, and how this knowledge may benefit the future of civilized society, we should hope that our association with the world of venoms will stay as intimate in the future as it has been in the past.

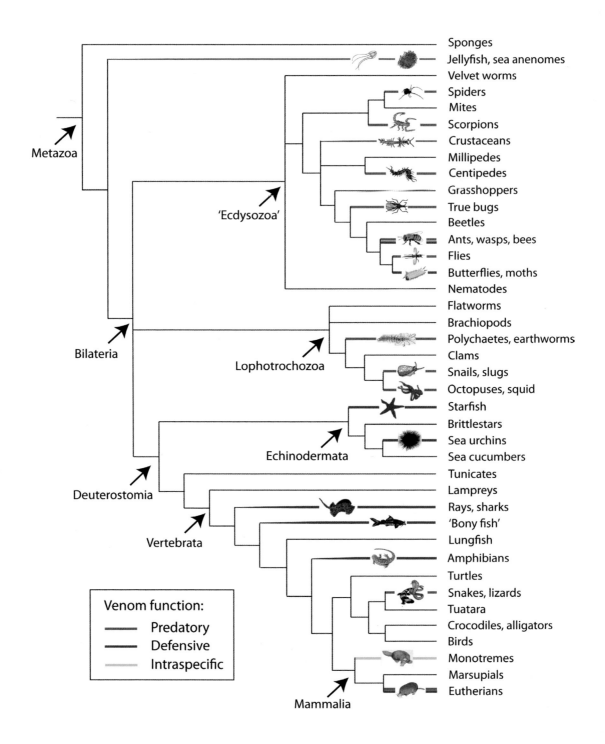

Sponges
Jellyfish, sea anenomes
Velvet worms
Spiders
Mites
Scorpions
Crustaceans
Millipedes
Centipedes
Grasshoppers
True bugs
Beetles
Ants, wasps, bees
Flies
Butterflies, moths
Nematodes
Flatworms
Brachiopods
Polychaetes, earthworms
Clams
Snails, slugs
Octopuses, squid
Starfish
Brittlestars
Sea urchins
Sea cucumbers
Tunicates
Lampreys
Rays, sharks
'Bony fish'
Lungfish
Amphibians
Turtles
Snakes, lizards
Tuatara
Crocodiles, alligators
Birds
Monotremes
Marsupials
Eutherians

Metazoa

'Ecdysozoa'

Bilateria

Lophotrochozoa

Echinodermata

Deuterostomia

Vertebrata

Mammalia

Venom function:
Predatory
Defensive
Intraspecific

Chapter 2

The Deadly Diversity of Venoms

The world of venom is astonishingly diverse. Consider the animal tree of life with its enormous diversity of body shapes and lifestyles. It surpasses even the imagination of the most creative fiction writers. Venomous animals occupy almost the entire range of possible shapes and sizes, from microscopic plankton to some of the largest animals on the planet. Venom plays a central role in the lives of most of these animals, but the number of different roles venom can have, and how venoms are delivered in order to fulfil these roles, stretches the imagination. This chapter provides a glimpse of this awesome diversity of venoms, the animals that harbour them, how venoms are delivered and what functions venoms have.

More than meets the eye

Our perception of venomous animals tends to be restricted to 'nasties', such as snakes, spiders, scorpions and sea nettles. This is not surprising, given the enormous impact many of these species have on the health and wellbeing of humans

An evolutionary tree of the animal kingdom that shows the relationships between the major groups of animals. Twenty lineages that contain venomous animals are indicated in colour, with the main functions of the venoms shown in different colours. This illustrates the broad distribution of venom in the animal kingdom, but venoms have evolved independently many more times (see pp.26–27).

This table estimates how many times venom has evolved in the animal kingdom, approximately how many species are venomous, and some of the main roles that venoms play in their lives. These estimates suggest that about one out of every eight described animal species is venomous. Two hotspots for the independent evolution of venoms are insects, which represent the most species-rich animal lineage, as well as ray-finned fishes.

Animal group	Venomous lineage	Times venom evolved	No. species	Main roles of the venom
Cnidarians				
	All	1	11,494	Predation, defence
Molluscs				
	Snails and slugs (Gastropoda)	6	7,748	Predation, defence
	Octopus, squid and relatives (Coleoidea)	1	806	Predation
Total		7	8,554	
Nemerteans				
	Venomous ribbon worms (Enopla)	1	724	Predation
Annelids				
	Blood worms (Glyceridae)	1	46	Predation
	Leeches (Hirudinea)	1	700	Blood feeding
Total		2	746	
Flatworms				
	Genus Mesostoma	1	150	Predation
Arthropods				
INSECTS	Flies (Diptera)	13	30,624	Blood feeding, predation
	Beetles (Coleoptera)	1	1	Defence
	Lice (Anoplura)	1	500	Blood feeding
	Fleas (Siphonaptera)	1	2,500	Blood feeding
	Lacewings, including antlions (Neuroptera)	1	6,000	Predation
	Wasps and relatives (Hymenoptera)	1	100,000	Defence, parasitoidism, predation
	Bugs (Heteroptera)	8	10,370	Predation, blood feeding
	Moths (Lepidoptera)	3	1,850	Defence
Total		29	151,845	

Animal group	Venomous lineage	Times venom evolved	No. species	Main roles of the venom
Arthropods (cont.)				
CRUSTACEANS	Remipedes (Remipeda)	1	28	Predation
	Carp lice (Branchiura)	1	210	Blood feeding
	Copepods (Copepoda)	2	2,178	Predation, blood feeding
	Skeleton shrimp (Caprellidae)	1	531	Intraspecific competition
	Isopods (Isopoda)	2	660	Blood feeding
Total		7	3,607	
MYRIAPODS	Centipedes (Chilopoda)	1	3,500	Predation, defence
ARACHNIDS	Pseudoscorpions (Pseudoscorpiones)	1	3,300	Predation
	Spiders (Araneae)	1	40,000	Predation, defence
	Ticks (Ixodida)	1	900	Blood feeding
	Scorpions (Scorpiones)	1	1,750	Predation, defence
Total		4	45,950	
Total		41	196,804	
Echinoderms				
	Sea stars (Asteroidea)	1	6	Defence
	Sea urchins (Echinoidea)	4	202	Defence
Total		5	208	
Chaetognaths				
	Arrow worms (Chaetognatha)	1	131	Predation
Chordates				
	Cartilaginous fish (Elasmobranchii)	3	212	Defence
	Ray-finned fish (Actinopterygii)	17	2,031	Defence
	Frogs and newts (Amphibia)	3	7	Defence
	Reptiles (Toxicofera)	1	4,600	Predation, defence
	Mammals (Mammalia)	5	446	Predation, blood feeding, intraspecific competition
Total		29	7,296	
TOTAL NUMBER VENOMOUS SPECIES		**88**	**226,107**	

worldwide, not to mention the innate fear many of us have towards them, learned over many generations. The same 'nasties' also make up the overwhelming majority of appearances of venomous animals in the media as well as scientific and fictional literature. Almost all coverage of venomous animals tends to relish the associated danger and fear while portraying the research and researchers as falling very close to the Indiana Jones end of the science spectrum.

However, these 'nasties' represent but a fraction of the immense diversity of venomous animals. This diversity is not just surprising, it is completely mind-blowing. Just consider a few examples: jellyfish graciously riding the ocean currents; glycerid bristle-worms digging their way through intertidal mudflats; sugar ants invading your cupboard at home; microscopic copepod crustaceans and chaetognath arrow worms hunting planktonic prey; colossal squid residing in the deep ocean; saw scaled vipers slithering across the Saharan desert sands; male platypuses patrolling Australian freshwater creeks; shrews foraging in the undergrowth. Over a quarter of all animal phyla are known to contain venomous species. A rough estimate of the number of known venomous species adds up to over 200,000, or almost 15% of all described animal species.

While general perceptions of venomous animals tend to be associated with fear, their diversity actually spans several aesthetic and emotive categories. They may be described as ugly, spectacularly beautiful or undeniably cute, with judgement resting very much in the eye of the beholder. Snakes of the viper genus *Atheris*, for example, are strikingly beautiful, although the fact that they are snakes disqualifies them from sympathy for many of us. Similarly, the peacock jumping spiders of southern Australia, *Maratus*, have achieved 'cute' status due to their brilliant colours and adorable mating dances. But again, spider cuteness is controversial. There are, however, animals about which there seems to be emotional consensus. For example, the Amazon toad fish, *Thalassophryne amazonica*, with its tissue-destroying venom, is far from aesthetically pleasing. Similarly, polychaete bristle-worms in the exclusively venomous Family Glyceridae, also often referred to as bloodworms, are not likely to top any ranking of the world's most beautiful organisms.

Male peacock jumping spiders try to attract females with elaborate courtship dances in which they wave their legs and bob and shake their colourful abdomens.

Nasty nature in a cute package

There are several venomous species that regularly turn up in forums showcasing beautiful animals. For example, the aptly named sea swallow, *Glaucus atlanticus*, is a truly beautiful and otherworldly looking creature. In fact, like several other nudibranch molluscs (sea slugs), the sea swallow does not actually produce its own venom, but harvests the stinging organelles from cnidarians that it feeds on. The theft of a complete venom apparatus from another species is exceedingly rare, and has only been documented in some sea slugs, a species of comb jelly, *Haeckelia rubra*, and several flatworms, all of them thieves of cnidarian stinging organelles. The stinging organelles, called nematocysts, are transferred through the gut of the sea swallow without discharging, and stored in the tips of tentacle-like or finger-shaped structures called cerata where they defend the creature against predators. Although the sea swallow feeds on several types of floating colonies of hydrozoan cnidarians it tends to store only the most toxic nematocysts it ingests, such as those from the Portuguese man-o-war, *Physalia physalis*. Moreover, the sea swallow stores only the larger of the two main types of nematocyst produced by *P. physalis*, probably because they possess the longest penetrating parts, which therefore provide the best defence when discharged. Thus, despite its innocuous name, the sea swallow is often considered even more venomous than the Portuguese man-o-war.

Venomous animals can also be so cute that it threatens their survival. Such is the sad fate of slow lorises, *Nycticebus* spp., which are the only venomous primates. Slow lorises inhabit the rainforests of Southeast Asia, and were considered fairly common until relatively recently when demand from the pet trade alongside habitat destruction have changed their International Union for Conservation of Nature Red List status to *Vulnerable* or *Endangered*. Slow lorises have a specialized gland on the inside of the elbow of their arms called a brachial gland that the animals use as both a poison and a venom (see p.13). Slow lorises anoint themselves and their infants with a mixture of saliva and brachial gland secretions, and experiments show that this mixture is not just toxic to ectoparasitic arthropods, but also deters smell-oriented predators such as leopards, tigers, sun bears and civets. In addition to using these toxins as a defensive poison, slow lorises deliver the mixture to opponents or predators via a venomous bite, which is powerful enough to result in severe symptoms including anaphylactic shock and even death in humans. As part of their threat display, slow lorises raise their arms over their head and lick their brachial glands while emitting loud hissing sounds. The display is not unlike the

THE DEADLY DIVERSITY OF VENOMS • 31

cobras that they are thought to mimic. Ironically, it is this cute-looking threat display that has become their greatest threat. People capture the animals to keep as pets, clipping their teeth to avoid a nasty bite and misinterpreting the animals' threat posture as an invitation to cuddle-time.

Sea swallows float upside down on the ocean surface, revealing their blue patterned bellies. They harvest nematocysts from the Portuguese man-o-war and sequester them in finger-like projections of their body wall, where they serve as powerful defensive weapons.

ALL SHAPES AND SIZES

Venomous animals are not just ugly, beautiful or cute, they also come in a truly astonishing range of shapes and sizes, including some of the smallest and largest species on the planet. For instance, predatory planktonic copepod crustaceans, such as those in the genus *Heterorhabdus*, at about 2 mm in length are clearly among the smallest known venomous animals. However, even they are relatively large compared to what are probably the smallest free-living venomous animals, the fairy wasps, a group of about 1,400 species of minute parasitoid wasps in the Family Mymaridae. Parasitoids are parasites that lay eggs on hosts or their eggs, which are then eaten and killed by the parasitoid larvae. Although fairy wasps in general are about 0.5–1 mm long, several species such as the aptly named *Tinkerbella nana* are less than 0.25 mm long. This makes them smaller than many single-celled organisms, and so small that their movement through the air is actually more comparable to swimming than flying.

While the smallest venomous animals are on the scale of the thickness of a human hair, the contenders for the largest venomous animal are among the largest animals known. The heaviest venomous species is undoubtedly the colossal squid, *Mesonychoteuthis hamiltoni*, which with a predicted maximum weight of roughly 750 kg (1,653 lb), is also the heaviest known invertebrate. It is more than 80 million times heavier than an adult *Tinkerbella*

A female of the tiny parasitoid fairy wasp, *Tinkerbella nana*.

nana, a relative size difference similar to that of a kitten compared to the world's largest aircraft carrier. However, with a maximum recorded length of about 6 m (19¾ ft), the colossal squid is far from the longest venomous animal. For example, a giant Pacific octopus, *Enteroctopus dofleini*, caught in British Colombia in 1956 had an arm span of 8.5 m (30 ft), while the longest giant squid, *Architeuthis dux*, was reportedly over 17 m (56 ft). Nevertheless, these are all dwarfed compared to the lion's mane jellyfish, *Cyanea* spp., with tentacles

that can span almost 37 m (121½ ft). This makes the lion's mane jellyfish almost seven times longer than the longest venomous snake, the king cobra, *Ophiophagus hannah*, and about 25,000 times longer than *Tinkerbella nana*.

The parrot-like beak of the colossal squid, *Mesonychoteuthis hamiltoni*. Cephalopods can inject venom into prey via bite wounds made with their beaks.

From drones to megabots

Appearances aside, the diversity of life forms found among venomous animals is truly astonishing. And in this sense, no venomous group is as diverse as the exclusively venomous cnidarians, which include jellyfish, sea anemones, corals and hydras. These range from free-swimming jellyfish to sessile (stuck on a substrate) polyps, and from solitary animals to colonial superorganisms. For example, although it superficially looks like one, a Portuguese man-o-war is no jellyfish. Instead it is a floating colony composed of seven specialized medusa-like and polyp-like parts, each performing a unique function. One of the polyp-like parts bears long tentacles that are richly endowed with venomous nematocysts, and these are used to catch squid and fish. Sometimes these tentacles are accidentally detached from the colony, and as they drift in the water they can end up stinging unsuspecting swimmers.

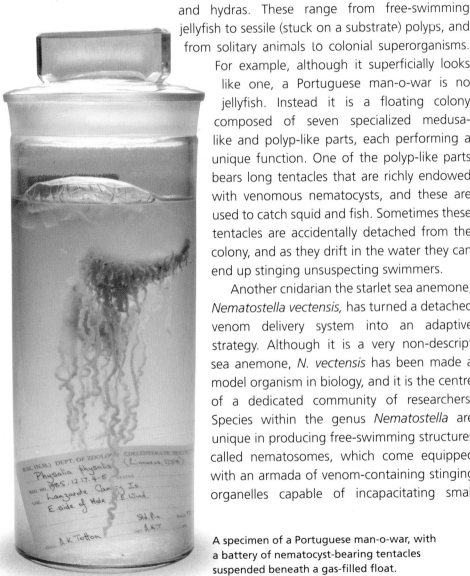

Another cnidarian the starlet sea anemone, *Nematostella vectensis,* has turned a detached venom delivery system into an adaptive strategy. Although it is a very non-descript sea anemone, *N. vectensis* has been made a model organism in biology, and it is the centre of a dedicated community of researchers. Species within the genus *Nematostella* are unique in producing free-swimming structures called nematosomes, which come equipped with an armada of venom-containing stinging organelles capable of incapacitating small

A specimen of a Portuguese man-o-war, with a battery of nematocyst-bearing tentacles suspended beneath a gas-filled float.

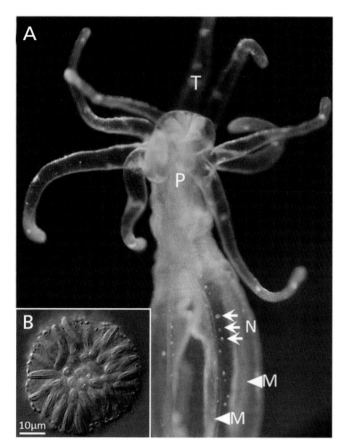

A polyp of *Nematostella vectensis* (A) with nematosomes (N) indicated by arrowheads. The pharynx (P) and tentacles (T) are also visible. (B) An optical section of a nematosome of *N. vectensis* with two types of stinging cells (cnidocytes or nematocytes) (b and m). Arrowheads indicate motile swimming cilia and arrows indicate sensory cilia that, when stimulated, trigger the discharge of the nematocysts.

crustacean prey. Nematosomes are generally contained in the body cavity (i.e. 'gut') of *Nematostella* where they are thought to help process ingested prey and contribute to the immune system. However, they are also incorporated into the jelly matrix of egg clusters, where they can protect the developing embryos from predators such as fish. Although they have cilia for locomotion, nematosomes are technically not independent organisms as they lack the ability to reproduce and regenerate. Instead, they are produced by folds of the gut called mesenteries, where they are loaded up with active stinging organelles before they are budded off in a sea anemone equivalent of a drone factory.

While nematosomes are technically speaking not venomous animals, there are cnidarians that remain extremely simple throughout their lives. Myxozoans are a truly bizarre group of cnidarians that have undergone a dramatic reduction in body complexity to become the perfect parasites, so much so that they were for a long time not even recognized as animals, and instead classified as protozoans (single-celled organisms). Their adaptation to a fully parasitic lifestyle has reduced their bodies to just a few cells, and they live as internal parasites of a variety of animals. Some species are even able to exist as intracellular parasites, for example infecting the egg cells or muscle cells of their hosts. Most myxozoans rely on two hosts to complete their life-cycle, and these tend to be a vertebrate (most often a fish) and an invertebrate (usually an annelid worm).

Although they consist of only a handful of cells, myxozoans have retained the cnidarian stinging apparatus, termed 'polar capsules' in myxozoans. Until recently, these polar capsules were thought to function as 'boarding' hooks that myxozoan spores used to physically attach to the host and initiate infection, rather than as venom delivery structures. However, detailed study of the myxozoan *Myxobolus* has revealed there are actually several types of polar capsules and some are able to deliver a secretion, or venom. Although the function of this venom remains to be determined, it does contain several toxin types that are found in other cnidarians. And at barely 5 microns across, myxozoans such as *Myxobolus* are certainly the smallest and most simplified venomous organisms on the planet.

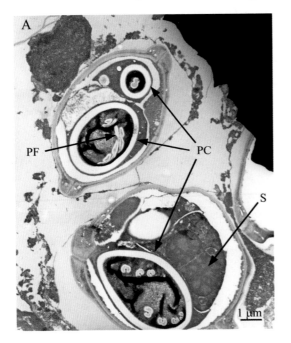

Two polar capsules (PC) with polar filaments (PF) of the myxozoan *Myxobolus stanlii* embedded in the connective tissue of the freshwater fish *Campostoma oligolepis*. Myxozoan polar capsules and polar filaments correspond to the nematocysts and the thread-like filaments of other cnidarians respectively.

VENOMOUS METROPOLISES

Besides representing the most simplified venomous animals as well as the most complex venomous colonies, some cnidarians are also the most impressive venomous ecosystem architects. Corals, like other cnidarians, catch their food with tentacles equipped with venomous nematocysts. Each coral reef formed by these tiny colonial relatives of sea anemones is a venomous metropolis that forms a home for an abundance of other organisms. And if a coral reef can be considered one large living organism, this would make the Great Barrier Reef of Australia by far the largest living organism on the planet, and the only venomous organism that is clearly visible from space.

The polyps of this Indo-Pacific orange tube coral, *Tubastrea faulkneri*, bear a crown of tentacles adorned with nematocysts, which they use to catch prey.

Loners and party animals

Besides representing a stunning diversity of life forms, venomous animals also encompass an incredible range of lifestyles. Among the venomous Hymenoptera (ants, bees and wasps), for example, species range from solitary individuals to superorganisms comprising millions of sisters ruled over by a single queen mother. The benefits of communal defence are obvious, and its impact can be devastating, especially when the defenders are armed with venom. It does not take much imagination to work out how Africanized honeybees got the name 'killer bees'. In contrast, the majority of hymenopterans that use their venom in prey capture are solitary hunters. Even the few social species that do hunt collectively often either do not use their venom for hunting, such as the paper wasp, *Parachartergus apicalis*, or no longer have functional venom systems. A peculiar example of the latter are the notorious driver ants of the genus *Dorylus*, which have lost their sting and repurposed their venom gland to make scent trails.

It remains a mystery why so few social venomous animals appear to use their venom during group hunting. Predatory venom has certainly evolved in most cases to enable more efficient solitary hunting. However, there are some notable and surprising cases of normally solitary hunters joining venomous forces. Spiders do not normally enjoy the company of other spiders even of the same species, yet different levels of social behaviour have evolved as many as 18 times in spiders. All of these species engage in group prey capture and feeding, and some even co-operate in everyday tasks such as nest maintenance and brood care. *Anelosimus eximius* is one such spider, and it can be found in colonies of about a thousand individuals. In scenes reminiscent of the cult classic *Arachnophobia*, numerous spiders swarm to any prey that ventures into their web, overpowering and killing it through collective envenomation. This allows this social spider to overpower prey 10–20 times bigger than itself. And the more the prey struggles, the more spiders come to the aid of their fellow attackers, joining in to cause death by a thousand venomous cuts.

Several pseudoscorpions (see p.196), a group also known as false scorpions, exhibit similar behaviour. Those of the genus *Paratemnoides*, for example, live in silken nests under tree bark in colonies consisting of individuals of both sexes and all ages. Although their 50-strong groups are much smaller than those of the spider *A. eximius*, they are truly social animals. Hunting is also a social occasion. For this, *Paratemnoides* individuals line up along the entrance of their colony to create a battery of venomous claws, and lie in wait for unsuspecting prey to pass. The unfortunate victim is seized by several adults, which then pull the prey into the colony and pin its body to the colony entrance. Like a big happy family hunt, the adolescents of the colony now join in, climbing onto the prey to inject venom through its soft joints. The victim of the hunt does not necessarily meet a swift end, however. Although a single pseudoscorpion has been observed to paralyze an ant in seconds, and kill it

A soldier of an African *Dorylus* driver ant stands ready for action. These ants use their jaws instead of their venom for predation and defence.

Anelosimus eximius spiders are social hunters. They trap prey, such as this grasshopper, in large communal webs, and dispatch them in group attacks.

within two minutes, it can take up to an hour after capture before the victim dies. Nevertheless, the strategy is clearly an effective one. Through social co-operation and venom use, pseudoscorpions are able to prey on other venomous species, such as ants and spiders, that are well over 30 times bigger. One species of *Paratemnoides* even goes to the extreme of self-sacrifice for the benefit of the group. When food is scarce a female *P. nidificator* offers herself to her nymphs. She remains motionless as she is consumed, after which the nymphs discard her empty exoskeleton from the nest. For the nymphs this communal matricide marks the start of a life as venomous pack hunters.

The diversity of envenomation methods

Considering the enormous diversity of venomous animals, it is unsurprising that they have evolved many different ways to deliver venom. After all, delivery of a cocktail of bioactive compounds is one of the defining features of being venomous. Stinging, piercing, drilling, scraping, biting, pinching, head-butting or spraying, the diversity of

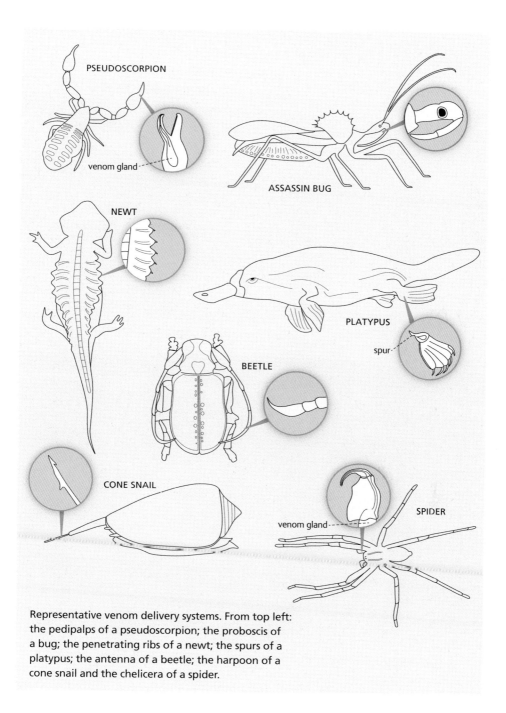

Representative venom delivery systems. From top left: the pedipalps of a pseudoscorpion; the proboscis of a bug; the penetrating ribs of a newt; the spurs of a platypus; the antenna of a beetle; the harpoon of a cone snail and the chelicera of a spider.

venom delivery mechanisms is truly impressive, as is the number of different structures that have been modified to deliver venom. These structures often show an extraordinary similarity in terms of their appearance and function, a topic that is explored in chapter 5. Most venomous organisms have evolved highly specialized and extremely efficient injection structures, which are even being used as models and sources of inspiration for research and development of human drug delivery equipment, such as syringes.

Evenomation by stingers

Stingers are probably the first type of venom delivery apparatus most of us think of, and for good reasons. The majority of venomous species on Earth inject venom in ways that can be described as stinging, and the specialized stinging structures known as nematocysts are even diagnostic for an entire animal phylum (Cnidaria). With a handful of exceptions, all venomous fish (bony and cartilaginous) deliver venom via the sting of more or less specialized spines that are often associated with fins or gill

The marbled stargazer *Uranoscopus bicinctus* has robust shoulder spines with associated venom glands. The one on its right side can be seen protruding from near the rear edge of its gill cover.

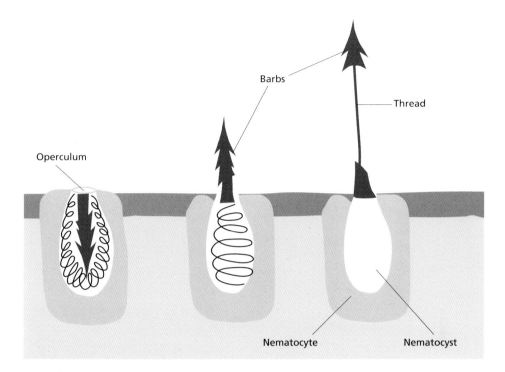

Barbs

Thread

Operculum

Nematocyte

Nematocyst

A schematic drawing of a cnidarian nematocyte with its nematocyst. The operculum in the left drawing is the lid that keeps the nematocyst closed. The two drawings on the right show the successive phases of nematocyst eversion after it is triggered.

covers. Some of these are hollow and syringe-like and have an elaborate venom-releasing mechanism, such as in toad fishes, *Thalassophryne* spp., and others are just grooved barbs lined with venom-secreting tissue, such as in the alien-looking rabbit fish, *Chimera monstrosa*. In insects, the modification of an ovipositor into a stinger can take much of the credit for Hymenoptera (bees, wasps, ants and their relatives) becoming one of the most species rich animal groups on the planet. A modified tail end has ensured the continuous survival of scorpions for over 400 million years, and the modification of the first pair of walking legs of centipedes into stinging appendages (forcipules) has allowed these to hang around for even longer.

Among the structures used for venom delivery those used for stinging or harpooning are among the most structurally specialized. Cnidarians sting with the gigantic secretory cell organelles called nematocysts, produced by specialized cells called nematocytes. Nematocysts are essentially high-pressure, venom-filled capsules that contain a harpoon-like stylet, and a tightly coiled, inverted syringe-like filament.

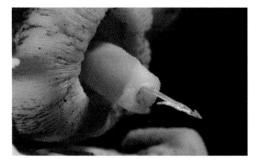

A harpoon-like radular tooth can be seen protruding from the proboscis of the fish-hunting cone snail, *Conus striatus*. This tooth delivers the snail's powerful paralytic venom, while securing the prey with the backwards pointing barbs.

Upon triggering, the stylet shoots out in the fastest known motion in the animal kingdom, accelerating from 0–100 kph (62 mph) in just 700 nanoseconds, with a piercing power similar to that of a bullet. Following penetration by the stylet, the filament is turned inside out in a twisting, drilling motion, reaching deep into the tissue of its victim where the toxic content of the nematocyst is released to cause paralysis and tissue destruction. This most complex of all animal cell organelles is also the oldest known venom delivery apparatus in the animal kingdom.

Another elaborate harpoon-like venom-injection mechanism is used by several members of the snail Superfamily Conoidea, which includes the venomous cone snails (Family Conidae), auger snails (Family Terebridae) and turrids (formerly classified together in Family Turridae). Here the venom is produced in a long duct-like venom gland equipped with a muscular bulb at the end furthest away from the opening. Contraction of the bulb flushes the venom through a hose-like proboscis terminating in a radula, which in most other gastropod molluscs is a tongue-like organ with teeth used for scraping up food. The radula of cone snails and their relatives, however, is equipped with a set of teeth that have become modified for delivering venom, either via stabbing or harpooning. Several conoideans have independently modified their radular teeth into elaborate harpoon-like hypodermic structures through which venom is injected. These hypodermic teeth are not actually attached to the radula, but are instead held tightly by sphincter muscles at the tip of the radula. The radular teeth of some species that feed on fish even feature a prominent barb that securely anchors their prey, allowing the snail to reel in their victim after injecting paralyzing venom. To ensure a steady supply of venom-injection equipment, the snail stores a number of these disposable hypodermic teeth in the radular sac, like a quiver of venomous darts.

Conoidean snails are not the only molluscs that have modified their radula into a venom-injection apparatus. Despite often being referred to as having a venomous bite, coleoid cephalopods (squid, cuttlefish and octopus) actually inject venom into prey and predators through a finger-like salivary papilla, although their beak may assist in causing an entry wound in soft-bodied victims. Strictly speaking, this salivary papilla is not part of the radula itself but is located right below, where it can also function as an accessory radula. The salivary papilla is connected to a pair of large venom-producing glands. Similar to conoidean snails, the end of the duct that connects the venom glands to the salivary papilla bears a set of teeth that aid in piercing prey during envenomation. However, these teeth are not as elaborate as the harpoon-like structures found in the advanced conoideans. The salivary papilla of octopuses, for instance, features a second set of teeth that allows them to drill though the armour of prey such as crabs and bivalve molluscs and inject paralyzing venom. During drilling, which can take up to several hours, the octopus also secretes a substance that dissolves the calcium carbonate of the shells. Once the shell has been pierced, the octopus injects venom that paralyzes the muscle that keeps the shell closed. This effectively turns the victim into a snack box and all that is left when the octopus is done eating is a shell with a single, millimetre-wide hole. Octopuses are important predators of shelled invertebrates, and have in some

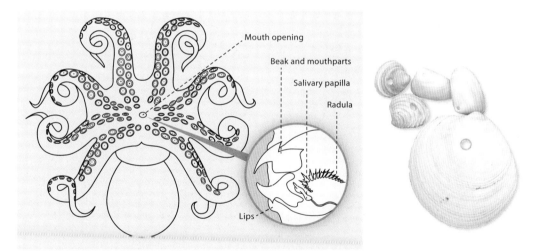

ABOVE The beak, radula, and toothed salivary papilla of an octopus can cause wounds through which the salivary papilla can inject venom into prey. The papilla is connected via a duct to the posterior salivary glands, which produce the venom.

ABOVE RIGHT Octopuses drilled holes in these bivalve shells so they could inject their venom, and eat the soft parts of their prey.

places been estimated to be responsible for over 90% of all predation on snails. Needless to say, drilled shells are commonly found washed up on beaches around the world, yet people may not realize that they are the evidence of a venomous attack assisted by mechano-chemical drilling.

Venomous bites

While strictly speaking cephalopods do not have a venomous bite, there are certainly many animals that do. The small pits often seen in dry sandy areas throughout the world are home to one such animal, the fierce and not-so-lovely-looking antlion larva. In fact, tens of thousands of insect species deliver venom through a bite, mostly as a means of predation during their larval stages. About 2,000 of these species belong to the antlion family (Myrmeleontidae). Antlion larvae have hollow mandibles which they use to inject paralyzing venom. The venom also liquefies the insides of insect prey, which the antlion then sucks back in through the hollow mandibles. This strategy is similar to that used by many other venomous insect

The semi-aquatic maggots of horseflies, like this *Tabanus imitans*, are predators that use their hollow, sickle-shaped jaws (visible in top right) to inject paralyzing venom into prey.

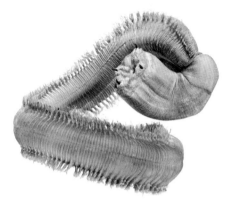

The marine bloodworm, *Glycera dibranchiata,* has four hollow teeth that it uses to inject prey with paralyzing and tissue liquifying venom.

larvae, such as those of horseflies (Family Tabanoidea). Although they may look like harmless maggots, horsefly larvae have a pair of hollow, sickle-shaped jaws at one end with which they are able to deliver venomous bites similar in pain to a wasp sting. Horsefly larvae are also voracious predators, feeding on anything from invertebrates to small amphibians, and they even cannibalize other horsefly larvae.

Although biting venomous maggots are an unpleasant surprise, they pale in comparison to another venomous worm-like creature. Venomous polychaete worms in the Family Glyceridae have a proboscis that they can protrude and which is equipped with four teeth squarely arranged in a truly alien-like fashion. These teeth are used for digging through the substrate of the intertidal flats they inhabit. To withstand the resulting constant wear and tear, the teeth have evolved into some of the hardest structures known to man. Reinforced with copper minerals, the teeth have an abrasion resistance approaching that of tooth enamel, and compete with the hardest metallic alloys. In addition to being effective digging tools, the teeth are also very effective at delivering venom. Each tooth is connected to an ample and muscular venom reservoir supplied with venom

SNAKES THAT STAB

The burrowing asps, *Atractaspis* spp., are an exception to delivering their venom via a bite though like other venomous reptiles, they still deliver the venom through their teeth. They use their long, highly mobile fangs to envenomate their victims through a sideways stabbing motion with their mouth closed. Also known as stiletto snakes and side-stabbing snakes, the burrowing asps appear to have evolved this peculiar behaviour as an adaptation to hunting subterranean prey in confined spaces. In such close quarters, striking from a distance like most snakes do is ineffective. Interestingly, they use just one fang during stabs, angling it out to the side while moving up alongside their prey before driving the fang deep into it with a sudden sideways movement of the head. This movement makes burrowing asps notoriously difficult to hold using the standard herpetologist's grip behind the head.

A drawing of the skull of the African burrowing asp, *Atractaspis aterrima*, showing the large fangs that deliver a powerful cardiotoxic venom.

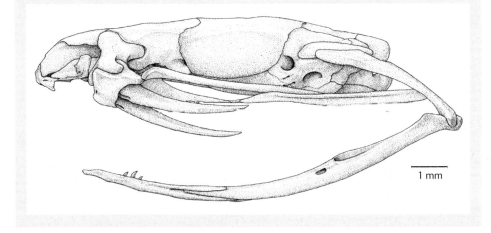

1 mm

from a set of lobe-like venom glands. Glycerids use their venom to subdue prey, which include crustaceans and other polychaete worms. Although the primary role of the venom is presumed to be predation, it is certainly capable of affecting humans. In the USA, where glycerids, also known as bloodworms, form the basis of a multimillion-dollar fishing-bait industry, painful bites are frequent and can result in serious allergic reactions.

Among vertebrates, most venomous mammals and reptiles use their teeth to deliver venom. These venom-delivering teeth show a wide range of specialization even within venomous snakes. For example, vipers inject pressurized venom through hollow, retractable, syringe-like fangs, while the Asian keelback snake, *Rhabdophis*

LEFT The hollow front fangs of the eastern diamondback rattlesnake, *Crotalus adamanteus*, deliver venom in a high pressure manner.

BELOW The boomslang, *Dispholidus typus*, delivers venom with large, grooved, rear fangs. It is one of the few snakes in the Family Colubridae that is dangerous to humans.

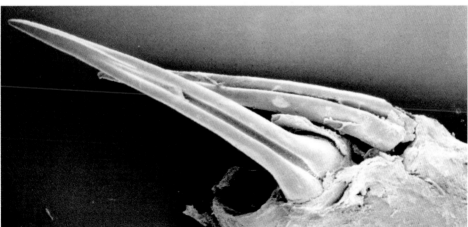

sp., employs a more passive diffusion into wounds created by non-grooved teeth. But the most common method of injecting venom used by snakes and other reptiles is via grooved fangs. In these cases, venom is drawn into the groove from the venom gland duct, due to its high viscosity and surface tension, and is essentially sucked into the tissue after penetration by the fang.

Hollow syringe-like fangs are found in only two genera of the Family Atractaspidinae (*Atractaspis* and *Homoroselaps*), vipers (Family Viperidae) and members of the cobra family (Elapidae). It's worth noting that the venomous snakes most deadly to humans all use such hollow fangs. These specialized fangs enable delivery of highly pressurized venom from muscular venom glands, which greatly increases the speed of delivery and volume of venom delivered. The Gaboon viper, *Bitis gabonica*, which with its 5 cm (2 in) fangs has the longest fangs of any snake, is able to deliver up to 600 mg of venom in a single bite. This represents over 40 times the total venom yield of some of the best yielding snakes with grooved fangs.

Spit, spray and secrete

In addition to the ability to rapidly deliver large quantities of venom, the development of hollow front fangs supported by a fixed upper jaw has also enabled the perhaps most infamous defensive venom delivery mechanism known among snakes, namely spitting. Spitting is thought to have evolved on several occasions within true cobras (*Naja* spp.), perhaps as a result of the spread of snake-killing hominids migrating out of Africa. Research suggests our lineage has always been one of enthusiastic snake killers. Venom spitting has also been documented outside true cobras, such as in the closely related rinkhals, *Hemachatus haemachatus*, and even in the spectacular Mangshan pit viper, *Protobothrops mangshanensis*. In all cases, spitting has become possible due to a narrower and more forward-pointing opening of the duct on the front tip of the fang. This increases the pressure of the venom system even further, resulting in a spit or spray that can reach up to 2 m (6½ ft), and which can be aimed with surprising accuracy.

Snakes are not the only animals that are known to spit or spray venom. For example, the red spotted assassin bug, *Platymeris rhadamanthus*, is a keen venom sprayer. It normally uses its proboscis to inject paralyzing and liquefying venom, produced in a pair of enormous venom glands that span from its thorax and

The African red spotted assassin bug, *Platymeris rhadamanthus,*
is an insect predator that can also spray its venom in defence.

far into its abdomen. It also secretes a smelly substance if stressed, but if this is
not enough to deter unwanted attention it will readily use its proboscis to spray
copious amounts of venom in a fashion not unlike an elephant showering itself
with its trunk. It can spray its venom over a distance of up to 30 cm (12 in), or
almost eight times its body length. Its venom also contains an abundance of
enzymes and other toxins that are able to induce intense local pain, relaxation of
blood vessels and swelling if the venom comes into contact with eyes or mucous
membranes.

Although spitting or spraying tends to be associated with defence, this is not
always the case. While readers may be familiar with acid-spraying ants, some ants
have taken this to a new level. The African ant *Crematogaster striatula* in particular
uses an astonishing approach to overpower and kill prey and deter competitors.
Instead of using their stingers directly, *C. striatula* ants line up in groups to surround
the prey at a safe distance, while aiming their stingers towards the victim. The ants

keep this position until after a few minutes the victim, such as a termite, appears intoxicated, keels over, and dies. Strangely enough, no visible spray is observed, suggesting that the active agents are volatile compounds. This raises the question of whether *C. striatula* ants are actually using venom at all, since no venom is actually delivered through a wound. The compounds thought to be responsible for this bizarre form of chemical attack are also not produced in the venom gland itself, but in an 'accessory gland' called the Dufour gland. In ants the Dufour gland opens at the base of the ovipositor or stinger where it can mix with venom from the venom gland, a process thought to be necessary for activating the insecticidal vapours of *Crematogaster* spp.

Ants in the genus *Crematogaster*, such as this *C. hesperus*, are sometimes called acrobat ants. They can bend their abdomens over their thorax and head and point it in almost any direction. Their stingers are spatulate and are used, not to sting, but to apply frothy venom like a brush, or simply to allow venom to evaporate, like in *C. striatula*.

This grey area between the definitions of poisons and venoms becomes even more diffuse in marine environments, where spitting or spraying venoms naturally becomes rather difficult. Aquatic animals nevertheless use various strategies to envenomate their victims from a distance. Sea anemones, for example, have toxin-secreting glandular cells that complement the action of their nematocysts, the venom-delivering stinging organelles characteristic of cnidarians. These glandular cells secrete potent neurotoxins, which may be absorbed through the gills of the unsuspecting victim, and aid the subsequent envenomation and capture by the anemone's nematocysts.

A similar tactic is employed by some cone snails, such as the deadly geography cone snail, *Conus geographus*, which secretes insulin-mimicking toxins that are absorbed across the gills of prey fish, causing them to enter a state of hypoglycaemic shock (see chapter 4). In a hunting strategy known as netting, up to several comatose fish are then engulfed by the snail's mouth, where envenomation by 'regular' harpooning ensures the fish do not wake up and escape. Unlike the neurotoxic peptides secreted by sea anemone glandular cells, the insulin-like peptides of *C. geographus* are produced in a bona fide venom gland. Using an insulin-like toxin to intoxicate prey is an example of the diverse actions venoms and their components can have, even in a seemingly straightforward function like predation.

More than just kill and defend

When considering possible functions of venom, the two that immediately spring to mind are predation and defence. These are undoubtedly the two most common drivers of venom evolution. Venom transforms the ecological battlefield between species engaged in the most direct struggle for survival from a physical to a chemical war. The evolutionary advantages of the use of venom for defence and predation are enormous, and underlie some of the greatest species radiations known in the animal kingdom. For example, there are well over 40,000 species of spiders and their enormous diversity can be almost exclusively attributed to the evolution of venom. It doesn't take much to imagine the alternative outcome of a plump soft-bodied animal armed with only small piercing mouthparts (spider) attacking a large insect with powerful spiny legs and crushing mandibles (grasshopper) without the assistance of venom. Similarly, any unsuspecting insectivore that comes across a giant silkworm moth

caterpillar, *Lonomia* spp., while looking for a tasty meal is in for a prickly surprise: the irritating hairs of *Lonomia* caterpillars contain venom that in some species, such as *L. obliqua*, is capable of inducing massive internal bleeding, destruction of blood cells, kidney failure and even death in humans.

However, venoms are more than chemical weapons used to incapacitate prey and deter predators. Even in predation and defence, they play a much wider range of roles than just killing and protecting. One of the most remarkable examples is found in an iconic venomous animal, the western diamondback rattlesnake, *Crotalus atrox*. Like most other vipers, *C. atrox* is a so-called strike-and-release predator, which means exactly what it says. However, its venom does not paralyze prey particularly quickly, with the result that *C. atrox* has to track down the location of its meal after biting it. This may not sound like an arduous task for an animal with a highly developed sense of smell and pit organs that can detect infrared radiation. Yet, the rattlesnake has to be able to distinguish the scent trail of the prey it has just envenomated from scent trails produced by other individuals of the same species that were not bitten. To solve this problem, the western diamondback uses its venom to relocate prey. Specifically, it uses two toxins called crotatroxin 1 and 2, disintegrin-type toxins that are present in the venoms of many vipers that normally disrupt blood clotting by inhibiting platelet aggregation. However, in *C. atrox*, and probably in several other pit vipers as well, the toxins also cause the release of volatile compounds through binding to a receptor in the prey. The resulting scent is a toxic homing device for the rattlesnake that allows it to relocate envenomated prey.

The scent of venom, however, can also attract unwanted attention to the venomous animal itself. Female parasitic flies in the genus *Pseudacteon* are attracted to the smell of venom components (alkaloids) and alarm pheromones produced by the fire ant *Solenopsis invicta*. Once she has chosen her target, the parasitic fly dive-bombs an ant and injects an egg into the ant's thorax in a fraction of a second. The egg hatches, and after about 10 days of eating the ant from the inside, the maggot moves into the ant's head and decapitates it, after which it pupates.

The venom of the western diamondback rattlesnake, *Crotalus atrox*, attacks the circulatory system of its prey, and causes internal bleeding. It also functions as a toxic homing device that allows the snake to track bitten prey.

ABOVE Attracted by the smell of ant venom, a female of the parasitoid phorid fly, *Pseudacteon tricuspis,* hovers near *Solenopsis invicta* fire ants, waiting for the opportunity to lay an egg on an ant in a fraction of a second.

LEFT A parasitoid phorid fly hatches from the decapitated head of a fire ant after eating the ant from the inside out.

Cooking with toxins

One of the characteristics of the venom of the western diamondback rattlesnake is that it is predominantly comprised of enzymes, which is why it is relatively slow-acting. Some scientists have speculated that enzymes in viper venoms may also play a role in aiding digestion, especially in the case of species that inhabit colder climates. However, regardless of the potentially digestive function of their venom, snakes are at least able to ingest their prey, a luxury not shared by all venomous animals. Predatory bugs, such as the common Australian assassin bug, *Pristhesancus plagipennis*, essentially eat their prey through a straw-like proboscis that they also use to inject their venom. This means that their venom needs to have components that liquefy their victims. These components come in the form of enzymes. Indeed, over half the components of *P. plagipennis* venom consist of type 1 serine proteases, which are enzymes that break down proteins. In addition to enzymes, *P. plagipennis* venom is also rich in pore-forming proteins, which contribute to liquefaction by breaking down cell membranes (chapter 4 discusses how pore-forming toxins work). These bugs literally melt their quarry with a 'kiss'.

While bugs use their venom to make a liquid meal, others use it to keep it liquid. Haematophagous animals, or blood suckers, need to keep the blood flowing in order to feed. They do this by injecting a cocktail of anticoagulants and vasodilators. Although readers may think it a stretch to call blood feeders such as mosquitoes, leeches, ticks and vampire bats venomous, the cocktail of compounds they deliver through their bites and stings do many of the things that 'classic' venoms do: they inhibit coagulation pathways and dilate blood vessels. This blood-sucking lifestyle is also best enjoyed without detection by the host, and this is achieved by the use of painkilling neurotoxins.

Preservatives added

While some animals use venom for processing their prey to facilitate ingestion, digestion, or both, others rely on venom for preserving their meals. Moles and shrews, for example, which devour insects and earthworms with razor sharp teeth, hardly need venom to overpower their prey. Instead, they use their neurotoxic venom for food storage, storing paralyzed prey in their burrows. Once paralyzed, the prey remains a fresh source of food for much longer than if it had been killed. For shrews in particular, storing food is essential to survival. Shrews have a fierce metabolic rate that is among the highest of any mammal. Despite mainly inhabiting cooler climates where food becomes relatively scarce during the winter months, shrews need to consume at least their own body weight in food every 24 hours. Needless to say, shrews would not last long without their venom-enabled fresh food supply.

Moles and shrews are not the only animals that use venom to store prey for later consumption. Some species have taken the preservative use of venom to another level. Venomous hymenopterans constitute the most species-rich group of venomous animals and include ants, bees and myriad wasp species. For many of these species, venom is not used for predation and arguably only secondarily for defence. Instead, venom is used to build safe habitats for their carnivorous but blind and largely helpless larvae by transforming animals into living meat larders. Among the most infamous of these is the tarantula hawk wasp, *Pepsis grossa*, a giant wasp known for having the second most painful insect sting after that of the bullet ant, *Paraponera clavata*. Like other spider wasps (Family Pompilidae), the female tarantula hawk wasp hunts large tarantulas that she paralyzes with a sting before dragging them to a pre-prepared burrow. Here she lays one egg on the spider's abdomen, before closing the burrow from the outside. Once the larva hatches from its egg, it immediately burrows in through the paralyzed abdomen of the still living spider and consumes it from the inside. Vital organs are left for last, maximising the time the larva has access to fresh meat.

This parasitoid lifestyle, which essentially consists of parasitizing the host to death, is very common among hymenopterans, and perhaps triggered the evolution of venom as a chemical aid to make the environment even safer for their larvae. Interestingly, parasitism stood at the evolutionary cradle of the stinging hymenopterans (Infraorder Aculeata), as all of them evolved from a common parasitoid ancestor. While many species use venom to 'simply' paralyze their larvae's hosts, this implies that the wasp needs to be able to manhandle the victim all the way to its nursery burrow. To get around this issue, the jewel wasp, *Ampulex compressa*, which targets cockroaches much too big for it to carry, has instead opted for creating zombies. Upon finding a suitable cockroach, the jewel wasp first injects venom into the thoracic ganglion to temporarily paralyze the front legs of its victim. With the legs temporarily disabled, she is able to place a second injection of venom directly into the brain of the cockroach. This second sting induces extensive grooming behaviour, then sluggishness and loss of normal escape responses. The jewel wasp then snips off about half of each antenna and periodically tastes the haemolymph (the body fluid of insects). Presumably by tasting the haemolymph the jewel wasp can check if she has injected an appropriate amount of venom, as too little could risk the cockroach recovering, while too much could kill it. Once satisfied with the result, the wasp leads her zombie victim to a hole where she lays a single egg on its abdomen before sealing off the tomb/nursery with pebbles. The hatched larva first feeds off the exterior of the cockroach before it burrows itself into its abdomen, where it lives its last stages as an endoparasitoid, feeding on the cockroach's organs while keeping it alive as long as possible.

The molecular mechanisms by which the jewel wasp venom zombifies her hosts are still being disentangled, but an eerie similarity has emerged between the state of zombified cockroaches and human patients with some neurological disorders. In the lethargic state that follows the cockroach's intense grooming, it has a dramatically reduced drive to self-initiate locomotion. The insect is not paralyzed, because it walks along as the wasp leads it to her nest burrow, but the cockroach seems to lack all motivation for moving. Intriguingly, this zombie-like state of hypoactivity is correlated with an insensitivity to the neurotransmitter dopamine, which suggests that something in the wasp's venom disrupts normal dopamine-dependent signalling in the cockroach's nervous system.

A jewel wasp has stung a cockroach and tries to wrestle it to her nest so she can lay her egg on it. The wasp's venom zombifies the cockroach by making it unable to initiate movement without being prodded.

A severe disruption of the dopaminergic system is likewise characteristic of people with severe Parkinson's disease, especially patients who survived the early 20th century sleeping-sickness (*encephalitis lethargica*) epidemic. As so vividly described by the late neurologist Oliver Sacks in his book *Awakenings*, post-encephalitic people would become frozen in place, unable to move. But like the envenomated cockroach they were not paralyzed. When presented with an appropriate external stimulus, they could move. For example, completely frozen patients could perfectly catch and return a ball thrown to them.

Venoms and habitat creation

Although parasitoids constitute a huge diversity of hymenopterans, not all hymenopteran larvae are carnivorous. Wood wasps such as *Sirex noctilio*, for example, are neither carnivorous nor parasitic during any part of their lives. Yet the larvae rely on the mother's venom for both creating a suitable habitat and a reliable food source. As their name implies, the larval wood wasps burrow in trees. *Sirex noctilio* is particularly fond of pines, and has become an economically important pest in the forestry industry in many areas of the world where it was accidentally introduced from its native range in Eurasia and northern Africa. However, the larvae do not actually feed on the tree itself, but on a symbiotic white-rot fungus, *Amylostereum areolatum*, that the female wasp injects into the tree together with her venom at the same time as she lays her eggs. *Sirex noctilio* venom weakens the tree's defence systems and allows the fungus to spread and grow, thereby creating the perfect environment for the wasp larvae.

Venom is a rare gardening tool. However, wood wasps are not the only animals that use venom to manipulate the growth of plants in order to improve their habitat. Like other ants in the sub-family Formicinae, the Amazonian lemon ant, *Myrmelachista schumanni,* compensates for the lack of a well-developed stinger with a large venom gland that produces lots of formic acid. Unlike other formicine ants, however, *M. schumanni* uses its venom as a herbicide to create the perfect habitat for its colonies' nest sites. It prefers trees of the species *Duroia hirsuta*, one of three tree species that live in mutualistic relationships with *M. schumanni*, and are called myrmecophytes (literally 'ant plants'). The lemon ants attack other plants by biting them and injecting formic acid, which kills them. This results in eerie mono-cultural patches of forest, which according to local superstitious beliefs are maintained by the evil forest spirit named Chullachaki. These forest patches were called Devil's gardens by Christian colonists.

The venom of the wood wasp, *Sirex noctilio,* aids the survival and feeding of its larvae in pine trees. The venom also causes pine needles to yellow and die.

Love thy neighbour?

Venom is not just used in the interaction between species, but also in competition between individuals of the same species. Some sea anemones provide the most extreme examples of this. As cnidarians, sea anemones are totally dependent on their venom for predation, defence and competition, and they have evolved specialized structures for each purpose. There are numerous types of nematocysts, with up to 30 types recognized across the Phylum Cnidaria. A single animal can have several types of nematocysts that perform specific roles. Some sea anemones, such as the clonal anemone, *Anthopleura elegantissima*, produce specialized nematocyst-filled sacs called acrorhagi that they use to compete with other sea anemones for space in crowded rock pools. Interestingly, acrorhagi contain different nematocysts compared to those on the tentacles and gut filaments, which are involved in prey capture and

Two *Anthopleura elegantissima* sea anemones engage in clonal warfare, stinging each other with their white-tipped acrorhagi. The more slender green tentacles are feeding tentacles.

digestion. Sea anemones also reproduce asexually, and are able to distinguish between clonal polyps (clones of themselves, regarded as friends) and non-clonal polyps (not clones of themselves and therefore enemies). Non-clonal polyps are viciously attacked with the acrorhagi, which detach from the attacker and attach to the wall of the victim, where nematocysts continue to fire and inject venom. These nematocysts also contain several toxins that are used specifically for competition against other anemones, and that cause severe necrosis and tissue destruction.

While sea anemones use their venom for predation, defence and intraspecific warfare, some animals use their venom solely on members of their own species. The platypus is a deceptively peaceful-looking mammal with its soft fur and cute beak, but males have a large spur on each of their hind feet, connected to a gland called the crural gland, which produces what is reputedly one of the most painful venoms known to man. Although this may sound like an excellent defence against predators – and platypuses do sometimes use it in defence, as we will see in the next chapter – only males have the ability to produce and deliver venom. Moreover,

Male platypuses are the only mammals that deliver venom by stabbing, which they do with a keratinous spur on their hindlegs.

venom production is cyclical, coinciding with the mating season. Platypus venom is therefore thought to be an extreme case of chemical male brawling, with a male-to-male generational arms race for territory and females that has resulted in the evolution of an incredibly painful venom. Interestingly, the males of the only other living egg-laying mammals, the echidnas, also have a spur connected to a crural gland on each hind leg that is only active during the breeding season. Unlike the platypus, however, echidnas cannot erect their spurs, and the crural glands don't have a venomous function. Instead, echidna crural secretions appear to be used for chemical communication rather than chemical violence. Although the idea is controversial, it has been suggested that the platypus and echidnas share an aquatic common ancestor that was possibly venomous, and that echidnas have a regressed venom system.

Venom love potions

Although male-to-male competition remains the most widely accepted hypothesis regarding the role of platypus venom, there are some scientists that believe it may have another, or at least an additional, role during the breeding season, namely preventing females from mating with additional males. The animal kingdom is full of examples of males trying to prevent females from mating with other males. Examples range from pure male-to-male combat, to the self-sacrifice of red-back spider males that block females' genitalia with their broken-off pedipalps, to males of the orb-weaving spider, *Larinia jeskovi,* that destroy external parts of the female's genitalia. The alternative hypothesis for the role of platypus venom is disturbing: if the male envenomates the female with extremely painful venom after mating he ensures that she will not mate with another male for at least another season. Though not confirmed, this could be another dark side to Australia's cutest darling mammal.

The role of platypus venom during mating remains a matter of debate, but there are animals that do appear to use venom for mating. Males of several scorpion species, for example, sting the females during courtship in a behaviour often referred to as sexual stinging. Scorpions do not mate directly, but instead have elaborate mating rituals that include the male grabbing the female by the pedipalps (the pair of large claws) and guiding her through a dance-like motion, termed the *promenade á deux,* to the position where he has deposited a sperm package. Although it remains unknown whether the stinging actually includes

The beautiful Indian tiger centipede, *Scolopendra hardwickei*, delivers powerful venomous bites to paralyze its prey, but males also bite females during mating.

envenomation, it does seem to have a calming effect on the female, allowing the male to guide her to pick up his sperm. In some scorpion species the male keeps his stinger inside the female for 20 minutes or more.

Like many scorpions, some centipedes also use sexual stings during courtship. Males of the gorgeous orange and black patterned Indian tiger centipede, *Scolopendra hardwickei*, for example, bite the females during mating, although the behavioural results are less clear than in scorpions. However, male *S. hardwickei* do have venom glands about 30% larger than those of females and they produce a suite of neurotoxin-like peptides that the females do not, which is a potentially significant investment in a love potion.

Topical toxins

Venom glands are perfect little incubators for bacteria and other microbes. They are accessible through a short tunnel that lacks physical barriers, they are full of nutritious protein, and are often inaccessible to the body's immune system. It is no wonder, therefore, that venoms tend to contain a number of compounds with potent anti-microbial activity. In most cases this anti-microbial activity can be considered a means of preserving venom prior to injection rather than contributing to the toxicity of the venom. This is in many ways similar to the activation of toxins, such as pore-forming proteins, that takes place upon envenomation. Many toxins can only be stored in their inactive state, and need to be modified to become active. However, there are examples where the venom plays a *bona fide* anti-microbial role after it has been ejected.

In competitive interactions with other ants the fire ant, *Solenopsis invicta*, smears or flicks its venom onto them. Many formicine ants, such as the raspberry crazy ant, can detoxify the fire ant's venom by applying their own venom to it.

The garden ant *Lasius neglectus*, for example, uses its venom to combat fungal infections. Like many other social insects, *L. neglectus* has an effective health-care system where peer grooming removes pathogens, such as fungal spores, from exposed individuals. However, *L. neglectus* also prevents the further growth of any overlooked spores by anointing each other with venom. Unlike the anti-microbial activities of most other venoms, which are usually due to minor components of the venom, this decontamination is largely the result of formic acid, which makes up over 60% of *L. neglectus* venom. Formic acid is also used as a cuticular ointment by other formicine ant species.

The curiously named raspberry crazy ant, *Nylanderia fulva*, is an invasive species in the USA. However, it is often considered the lesser of two evils as it actively displaces another invasive species, the notorious and destructive fire ant, *Solenopsis invicta*. As its common name suggests, the fire ant is armed with a powerful venom that tends to scare off or kill any competitors. However, upon contact with fire ant venom, the raspberry crazy ant quickly applies some of its own venom by self-grooming, and thereby neutralizes the fire ant venom. Although the raspberry crazy ant shows a similar behaviour with other ants, the neutralizing effect of its antivenom is strongest against the fire ant venom, probably reflecting their native geographical overlap and previous co-evolution. It demonstrates an arms race in the most classic sense.

There is one particular situation in which formicine ants often fail to deploy their acid weapon, with fatal consequences. Antlion larvae often feed on them, and in most cases the ants are killed without even attempting to spray their formic acid onto the antlion grub. If they did this, their chances of survival would be markedly improved because antlions are sensitive to formic acid. Experiments show that when a small quantity of formic acid, equivalent to the amount a single ant can deliver, is applied directly to an antlion, it will, without fail, release its prey in seconds. Tragically, however, formicine ants, such as *Camponotus floridanus*, typically only spray their acid when they can first get a purchase on the enemy with their mandibles. If they can't bite, they don't spray. By grabbing the ant in such a way that it is prevented from biting, the antlion manages to disable the ant's toxic weapon.

This chapter shows that diversity is *the* hallmark of the venomous world. No matter which dimension of this world you choose to explore – the variety of species, the habitats where they live, their range of venom-delivery structures, mechanisms and behaviours, the roles that venom plays in their lives – diversity rules. The next two chapters will explore the fundamental pillar that underpins all this variation: the venom cocktail.

Chapter 3

Probing the Power of Venom

One night in April 2012, biologist George Madani was spotlighting in the jungle of Sarawak, searching for wildlife. He noticed a slow loris, *Nycticebus kayan*, about 2 m (6½ ft) up a mango tree. To get closer Madani climbed the tree, which caused the little primate to fall to the ground. When he picked it up the loris raised its arms above its head, as if to invite him to tickle it. In reality, this is a defensive posture. The loris bit Madani on his finger, holding on for 30 seconds. Within two minutes of the bite Madani felt a tingling sensation in his jaw, ear and right foot. Within an hour his face had deformed, with his lips swelling cartoonishly. When he arrived in hospital he was suffering from anaphylactic shock. He was agitated, nauseous, short of breath, felt weak and had chest pains.

Snakes have bitten our colleague Bryan Fry 27 times, so far. The scariest of these experiences was the first time he tried to milk venom from a Stephen's banded snake,

OPPOSITE A slow loris in Borneo.

RIGHT AND FAR RIGHT Biologist George Madani 54 minutes after he was bitten by a slow loris, showing pronounced facial swelling. George Madani a week after the bite, after the anaphylaxis symptoms had disappeared.

Hoplocephalus stephensii. Fry had collected the snake not far from the University of Queensland, Australia, where he was working as a PhD student studying snake venom. The snake bit him on his right index finger, which immediately resulted in a pounding headache and a crushing feeling on his chest. Before long Fry lost consciousness. After coming to, he was rushed to hospital where for the next 30 hours he was treated for what medical parlance ominously refers to as venom-induced consumption coagulopathy. The snake's venom had destroyed Fry's blood-clotting ability by activating and thereby depleting all his clotting factors. In a small animal the clots induced by the venom quickly lead to stroke injury and death. But in the larger blood volume of a human the clots are diluted enough not to cause immediate damage. The tiny blood clots coursed through Fry's body, threatening to block vital blood vessels that could lead to stroke and cardiac arrest while the depletion of all his clotting factors caused him to bleed from eyes to anus. Luckily, he survived.

In June 2016, one of the authors was holidaying in Taman Negara National Park in Malaysia. This tropical rainforest teems with wildlife, including tigers, elephants and snakes. One morning Ronald woke up to discover that he had been assaulted by a venomous animal during the night, which left him with a swollen wound on his leg. He quickly applied some cream to the mosquito bite and went to breakfast.

In May 1991, a 57-year-old war veteran and Victoria Cross holder was fishing at a mountain resort in North Queensland. He saw a small platypus floating nearby, and thinking the animal might be sick or injured, he decided to pick it up by the scruff of its neck. But when he tried to put the animal back in the water, it drove two keratinous spurs on its hind legs into the man's right hand. The pain was immediate and brutal, and developed over several hours into severe whole body pain. Morphine infusions barely relieved the agony. Only a chemical block of the nerves to the envenomated hand made the pain more bearable. Three days after the sting the veteran's body pain had finally subsided, and the pain in his hand had become tolerable. A month after being stung his hand was still painful, and after three months his fingers and hand were still swollen.

On 11 September, 2001, California Academy of Sciences herpetologist Joe Slowinski was on fieldwork in the jungles of Myanmar. Before breakfast he was handed a bag by a field assistant, who said it contained a wolf snake, from the genus *Dinodon*. These pretty black-and-white banded snakes are harmless. But when Slowinski retrieved his hand from the bag, a pencil-thin, foot-long snake dangled from his middle finger. 'That's a fucking krait!' he said. He had been bitten by a multi-banded krait, *Bungarus multicinctus*, similar in appearance to the wolf snake but much more dangerous. The bite was painless and left no mark. Slowinski proceeded to have breakfast, before he lay down for a nap. Then the muscles in his hand started to tingle. Over the next few hours Slowinski slowly slipped into complete paralysis. Cruelly, he remained conscious throughout this torturous ordeal. His companions gave him mouth-to-mouth for 26 hours. After his pulse could no longer be detected they gave him 3 hours of CPR, but to no avail. Joe Slowinski was dead.

When human lives intersect with those of venomous animals, the outcomes fall along a broad spectrum from trivial to traumatic, from uneventful to fatal. Cases such as these provide scientists with important insights into how venoms attack bodies. We know a lot about the terrible power of venom, especially snake venom, because human envenomations often require medical treatment. Snakebites can cause horrific syndromes of symptoms, with single bites disrupting multiple organ systems, and with symptoms worsening over a period of days if the venom is not neutralized in time with antivenom. Envenomations by vipers and spitting cobras can be particularly destructive, causing widespread tissue damage that can lead to permanent disability. But our understanding of how venoms attack is not only

Banded sea kraits, like this *Laticauda colubrina,* have powerful paralytic venoms, but their bites are often fairly painless and may leave no discernable marks on the skin. In Southeast Asia kraits frequently bite sleeping people without waking them up.

based on accidents. Natural history observations have contributed many insights into the menacing mechanics of venoms as well. For the observant naturalist, a wave of necrosis spreading along the vein of a leaf in the Amazonian jungle betrays a formic acid envenomation by the plant-killing lemon ant, *Myrmelachista schumanni*. And a diver who observes a marine snail protruding conspicuously from its shell, with its foot contracting in violent spasms, knows that the snail has likely been envenomated by the mollusc-hunting cone snail, *Conus textile*. But by far the most detailed insights into the power of venom come from scientific experiments.

In this chapter and the next we will review what scientists have learned about the power of venom, and how it makes venom both a force for good and for evil. This will allow us to return to the above case studies and understand how and why each of the venoms in question managed to cause the observed symptoms.

The unfortunate lab rat

To be able to prevent or neutralize the damage done by venoms, researchers need to understand what havoc they can wreak in a victim's body. Our deepest insights come from experimentally exposing animals to venoms. To reveal venoms' toxic secrets, toxinologists have exposed a menagerie of animals, including cats, dogs, rabbits, rats, mice, hamsters, guinea pigs, sheep, lizards, frogs, toads, chickens, clams, fish, insects, nematodes, monkeys and more, to a vast number of venoms and isolated toxins. The result is a toxinological literature filled with experiments of a macabre ingenuity that reveals the biological arsenal that Mother Nature has evolved to disrupt and destroy her own creations.

One such article is titled 'Fatal intoxication of rabbits, sheep and monkeys by the venom of the sea wasp (*Chironex fleckeri*)' from a 1972 issue of the journal *Toxicon*. It packs a large amount of misery in two short pages of dispassionate prose. A meeting abstract published in the same journal and year investigated the question of whether bites by moray eels are venomous. The single sentence that describes the authors' experimental results leaves both little and a lot to the imagination: '…bites of freshly caught moray eels showed a strong traumatic effect, but not a toxic one, in guinea pigs.' Another abstract in the same issue further captures the brutal reality of being a guinea pig in a toxinology lab. It describes how guinea pigs forced to inhale an aerosolized solution of assassin bug venom developed difficulty breathing within a minute, and after a short period of convulsions died by respiratory arrest.

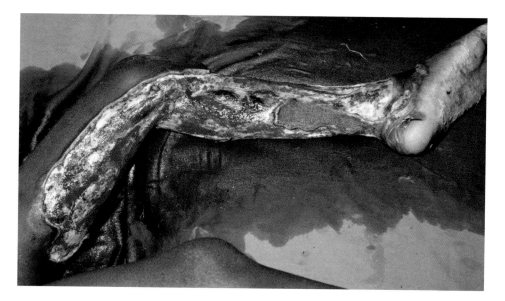

Catastrophic dermonecrosis (death of skin cells) affecting the
leg of a Kenyan girl, three weeks after she was bitten by Ashe's
spitting cobra, *Naja ashei*.

Although research like this makes for very grim reading, the experiments are not performed for frivolous reasons. Sea wasps are deadly jellyfish that envenomate many people every year, moray eels occasionally deliver serious bites to divers, and the species of assassin bug investigated above can kill a mouse in 30 seconds and is even rumoured to be responsible for several human deaths. It would be nice to think that such crude animal-based experiments into the effects of venoms are largely a thing of the past, but they aren't. Animal experiments remain a source of invaluable information about the physiological effects of venoms and toxins, and they provide an important framework for working towards more effective treatments of human envenomations.

Lethal power is therefore still routinely measured by injecting venoms or toxins into the skin, muscle, veins or body cavity of mice, and cardiotoxicity is often assessed with injections into the hearts of rodents or toads. Unfortunately, the suffering and death of some experimental animals is unavoidable if we want to understand the destructive power of venom. If scientists want to understand the potential effects of a rattlesnake envenomation on the functioning of the heart, so that snakebite victims can receive the best possible treatment, they have little choice but to inject

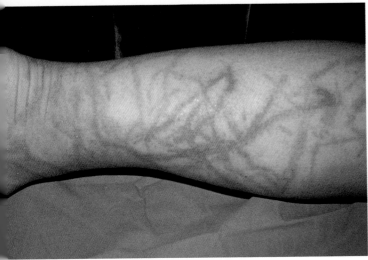

ABOVE The sea wasp, *Chironex fleckeri*, is the deadliest jellyfish in the world. Its devastating venom attacks the human body on many fronts, damaging the skin, nervous system, blood, heart, and muscles.

LEFT The leg of ten year old Rachel Shardlow who, in 2010, received near-fatal sea wasp stings when she was swimming in the Calliope River in Queensland, Australia.

rattlesnake venom into experimental animals to see how it causes myocardial infarction. And to understand the effects of stepping on the painfully venomous spines of a greater weever fish, *Trachinus draco*, without doing this themselves, researchers can hardly avoid injecting the venom into an experimental animal. In mice, this reveals that the venom causes tissue necrosis as a result of the contraction of blood vessels, which causes the animals to go bald, and to lose their ears and the tips of their tails.

Among the rationales behind the use of experimental animals to gauge the effects of venoms is that no single experiment with a cell culture or a tube of blood can reproduce the complex sets of effects that venoms have on living organisms. This is also the case for the study of toxins that act on perhaps the most complex organ of all, the nervous system. To identify pain-causing toxins, or toxins that can suppress pain, toxinologists inject venoms into the soles of mice, and then score the time the animals spend licking their paws as a measure of pain intensity. Similarly, to see if a venom or toxin can have an effect on the central nervous system, they are injected into the brains of mice, and the behavioural consequences recorded. And to investigate if a venom can affect an animal's sensory organs, it can be added to water to see if it affects the swimming behaviour of larvae of the zebrafish, *Danio rerio*.

While animal-based experiments remain crucial for many aspects of toxinological research, experiments using live animals are subject to tight control and strict regulations. Every possible measure is taken to minimize the suffering of lab animals. For instance, to gauge the muscle-destroying and necrotic potential of venoms they are injected into the calf muscles of mice, but the animals are first anaesthetized whenever possible. Similarly, to see if a venom or toxin affects blood pressure, it can be injected into anaesthetized rats or cats.

Zebrafish larvae are used by toxinologists in experiments to assess the effects of venom toxins on their behaviour and body.

Moreover, many investigations into the effects of venoms and toxins do not require live animals. Scientists have invented a range of experiments using nerves and tissues dissected from lab animals to test the activity of toxins. For example, to see how toxins affect the contraction of smooth muscles, toxinologists bathe the gut of a guinea pig, or the uterus or vas deferens of a rat, in a solution to which a venom or toxin has been added. A classical neurophysiological tool for studying the transmission of nerve impulses to muscles is technically known as the 'isolated chick biventer cervicis nerve-muscle preparation'. Remarkably, about half of all scientific publications that use this preparation of a chick's neck muscle are concerned with testing the neurotoxic effects of snake venoms and toxins. Of course experiments such as these still require the sacrifice of animals, but their suffering is minimized.

These days, one of the most common approaches to testing the activity of toxins is to measure their effect on cells that contain the targets of interest. To discover neurotoxins and to study how they work, toxinologists can, for example, use egg cells of the African tongue-less clawed frog, *Xenopus laevis*. They can test whether the toxins affect specific proteins involved in the transmission of nerve

Egg cells of the African clawed frog, *Xenopus laevis*, are used as a system to study the effects of neurotoxins present in animals' venoms.

impulses, such as neurotransmitter receptors or ion channels, by injecting the genetic precursors (known as messenger RNA molecules) of these target proteins into the egg cells, which then go on to produce the receptors and ion channels. By injecting the human precursors for these proteins into the frog egg cells, researchers can even test whether humans would be sensitive to the neurotoxins. To assess sensitivity, researchers measure whether addition of a venom or toxin to the solution in which the egg cells are incubated causes changes in the current of ions across the cell membrane of the egg cell. Such experiments are supplemented by assays of enzyme activity, anti-microbial activity, blood clotting and blood cell toxicity, among others. Indeed, given the enormous diversity of roles and functions of venoms, there is virtually no experiment that cannot produce a fact that can be woven into the growing tapestry of toxinological knowledge.

What's in the cocktails?

The key to understanding the symptoms caused by envenomations is the toxin composition of venoms. How do scientists determine the complexity of these toxic cocktails? How do they obtain venom in the first place? And how do scientists measure the lethal power of venom with a common standard? These are the three questions we will address in the remainder of this chapter.

Venom is a gloopy, salty broth. It may look homogeneous in a test tube, but it contains a miscellany of chemicals. Venom typically contains a complex mixture of salts, peptides and proteins – such as enzymes – as well as other organic components such as amino acids, lipids and amines. Amines are a group of compounds that includes neurotransmitters such as dopamine and serotonin. As a general rule, peptides and proteins are the most abundant as well as functionally the most important ingredients of venoms, although there are some fascinating exceptions to this rule, such as the formic acid-producing venom glands of formicine ants (see below and chapter 2). The high protein concentration makes many venoms quite viscous. This is obvious when collecting venom from, for instance, marine glycerid worms. When a worm's dissected venom glands are squeezed with forceps, thick globules of venom appear on the tips of the worm's jaws. The globules stick to the jaws so well that it takes vigorous rinsing to remove them.

Much venom research concentrates on characterizing the mixtures of peptides and proteins present in venoms. Peptides and proteins are organic molecules that are made up of strings of amino acids. There are 20 different amino acids, but not all species can produce each of them. Humans, for instance, must obtain nine of them, known as the essential amino acids, from food. Strings of amino acids form peptides and proteins analogous to the way that letters add up to form words. Each unique sequence of amino acids makes up a unique protein or peptide. Whether these natural polymers are called peptides or proteins principally depends on their size. Generally, a peptide is 100 or less amino acids long, while anything larger is called a protein, although different authors draw the line in different places. Protein is also used as a general term for any amino acid chain independent of its length. In this book we use protein in this general sense, unless we use it to directly contrast it with peptides, in which case it refers to longer amino acid chains.

Venom cocktails typically contain a mixture of peptides and proteins, with their relative abundance varying among the different venomous animals. The venom of a single front-fanged snake, such as a mamba or a rattlesnake, may contain over a hundred proteins representing a dozen or so different types of toxins. The most complex venoms, such as those of cone snails and spiders, can contain hundreds or even thousands of different proteins. Remarkably, there may be no overlap at all in the set of proteins present in the venoms of even closely related species, a situation that is especially striking for the complex venoms of cone snails. Estimates of the total diversity of proteins that can be found in animal venoms are therefore almost astronomical. For instance, the venoms of spiders alone – a group containing more than 40,000 species – may represent a diversity of over 10 million distinct proteins.

As a rough rule, venoms that have evolved primarily for the purpose of predation, like those of spiders, snakes, scorpions and cone snails, tend to be more complex than venoms that are purely or mainly for defence, such as those of fish, bees and many ants. This is in part because the main function of defensive venoms is to cause pain, which can be achieved more easily than causing the physiological breakdown needed to disable a prey animal. The very simplest venoms, such as those of the formicine ants, which lack a functional stinger, may be dominated by just a single ingredient. Formicine ant venom contains almost no proteins, and consists mainly of formic acid, a corrosive and volatile liquid that the ants squirt at enemies.

However, there are some predatory venoms that are also relatively simple. The venoms of sea snakes and sea kraits – snakes from the cobra family (Elapidae), which have independently invaded the sea – are streamlined compared with their land-living relatives, likely as a consequence of them specializing on just a single prey type: fish.

Unmixing a venom cocktail

One of the most striking examples of fractionating a venom cocktail has come from the laboratory of Baldomero Olivera at the University of Utah. Even though the lab is 1,000 km (620 miles) away from the ocean, it has been an important testing ground for the venoms of predatory marine cone snails since the 1980s. When Olivera and his colleagues fractionated cone snail venom with liquid chromatography they revealed a very complex toxin mixture. But the most stunning result was what happened when they injected each venom fraction into the brains of mice. Each venom fraction of the dangerous fish-hunting cone snail, *Conus geographus*, elicited a very specific response. One fraction put mice to sleep, another made them swing their heads back and forth, while another produced twisted jumping. Yet other fractions made mice comatose, unco-ordinated, or caused trembling, convulsions and incessant scratching, or made them walk in circles or lie on their backs while kicking their legs in the air. Imagine what would happen if all these venom fractions were injected simultaneously. Whether you're a fish or a human, the result would be the same: death by paralysis. About 70% of people stung by a geography cone snail die if they don't receive medical care.

The deadliest snail in the world: the marine, fish-hunting, geography cone snail, *Conus geographus*. It manages to get close and even engulf its prey by triggering hypoglycaemic shock in fish before it stings them (see chapter four).

UNMIXING A VENOM COCKTAIL

The secret of any cocktail is its mix of ingredients. This is as true in a venom research laboratory as it is in a bar. But whereas even the most talented bartender could never unmix his own concoctions, venom researchers do this routinely. The process of unmixing a venom cocktail is called venom fractionation, and this can be achieved in several ways. Researchers take advantage of the fact that proteins differ in size, shape and electric charge. One method, known as gel electrophoresis, involves forcing venom to move through a gel placed in an electric field. This separates the venom components according to their mass and their electric charge, as they migrate through the gel at different speeds. The components can then be cut from the gel and analyzed in order to determine their identity. Another method, known as liquid chromatography, can separate venom components based on how quickly they pass through a column packed with a matrix composed of particles that the venom components stick to. By gradually changing the solvents that flow through the column, the venom components detach from the matrix and migrate through the column at different times depending on the electric charges or water-repelling properties on their surface. The venom fractions are collected in separate tubes as they exit the column, and can then be analyzed.

BELOW LEFT AND RIGHT: The venom of the bloodworm *Glycera dibranchiata* separated into its components by gel electrophoresis, and a corresponding gel of the venom of the toad fish, *Thalassophryne amazonica*. The defensive fish venom is much less complex than the predatory worm venom.

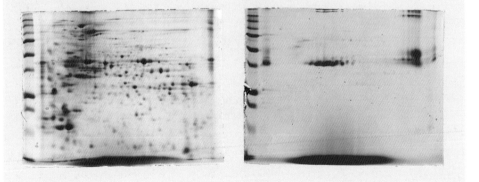

In order to understand how individual venom proteins act on their targets, you need to know their amino acid sequence. To identify the amino acid sequence of proteins scientists may use a technique known as mass spectrometry. The venom proteins are typically first fragmented, and these smaller peptides are then ionized in the mass spectrometer, which means they become positively or negatively charged ions. The ions are then separated on the basis of their mass and charge, creating for each a unique mass to charge ratio fingerprint. By comparing these fingerprints to existing databases of the known mass to charge ratios of proteins, scientists can infer the likely amino acid sequence of their sample. Scientists use these amino acid sequences of venom proteins to gain insights into their modes of action, possible biological roles and their evolutionary history.

Other techniques, such as nuclear magnetic resonance, can be used to determine the three-dimensional structure of proteins. The interactions between the amino acids in a protein determine how it folds in space, and a protein's folding in turn determines how it functions by specifying how and how strongly it can interact with other molecules, such as an enzyme and its substrate. But before scientists can study how venom proteins function inside the body of an envenomated victim, they first need to collect venom.

The art of venom milking

How do scientists coax animals to part from their precious venoms? Some supply it readily enough. Just annoy a spitting cobra and it will deliver its defensive secretion in two powerful jets from its fangs. Threaten the wrong kind of assassin bug and it, too, will squirt its venom, this time from its nozzle-like proboscis. Agitate a Sydney funnel-web spider, *Atrax robustus*, and venom will start dripping from its fangs, ready to be collected. It turns out that getting an octopus to squirt its venom is almost just as easy, although neatly collecting it for scientific work is another matter: put the animal in a clear plastic bag, gently prod it and roll it around, and wait for it to bite through the bag with its parrot beak-like jaws, evert its salivary papilla and squirt its toxic saliva into the air. The difficulty is catching it. Other animals, however, are more finicky, and require a more intimate approach.

Many readers will be familiar with photos or footage of researchers milking venom from snakes. Front-fanged snakes, such as cobras and rattlesnakes, can be milked by inducing them to bite into a container covered with an artificial membrane, such as parafilm (a plastic paraffin film). Gently massaging and squeezing the venom glands while the snake bites into the container can increase venom yield. Non-front fanged snakes, such as the African boomslang, often need to be sedated before milking to be able to access the venom fangs deep in the snake's mouth. A syringe-like pipette can then be placed over the fangs to aspirate

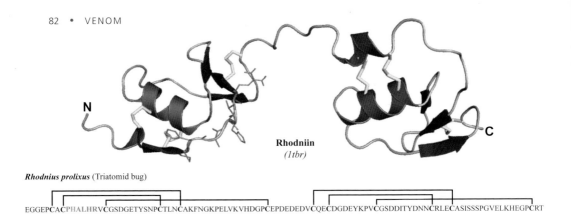

Rhodniin
(1tbr)

Rhodnius prolixus (Triatomid bug)

EGGEPCACPHALHRVCGSDGETYSNPCTLNCAKFNGKPELVKVHDGPCEPDEDEDVCQECDGDEYKPVCGSDDITYDNNCRLECASISSSPGVELKHEGPCRT

ABOVE A 3-D model of the protein rhodniin contained in the venom of the blood-sucking kissing bug, *Rhodnius prolixus*, with its corresponding amino acid sequence. The key bonds which confer stability – an important feature of venom toxins, see p.137 – are between the pairs of the amino acid cysteine (C) (indicated by brackets above the amino acid sequence and in yellow in the model). The two ends of the protein are indicated by N and C. Rhodniin, in the venom of this blood-feeding insect, prevents the action of the enzyme thrombin that is crucial for blood clotting, so allowing the insect to feed continuously on the blood of its prey.

OPPOSITE TOP Paul Rowley (left) and Nick Casewell of the Liverpool School of Tropical Medicine, UK are milking venom from a subadult Gaboon viper, *Bitis gabonica*.

OPPOSITE BOTTOM The eastern brown snake, *Pseudonaja textilis*, is one of the most venomous snakes on Earth. Milking venom from its short, stiff fangs with a pipette requires a considerable amount of dexterity.

the venom. The same technique is used to collect venom from monitor lizards and their relatives. Substantial amounts of venom can quite easily be milked from the high-pressure venom delivery systems of front-fanged snakes, but it can be difficult to obtain sufficient quantities of venom from non-front fanged snakes and lizards. To increase venom yield in these cases researchers use a trick.

To stimulate venom secretion they inject a plant-derived alkaloid, called pilocarpine, near the venom glands. Pilocarpine causes salivation in many animals, making it a useful drug for stimulating venom secretion in very different animals. Vampire bats can be anesthetized and then injected with pilocarpine to stimulate the secretion of their toxic, anticoagulant venom. The same works well in giant predatory waterbugs, and when a drop of pilocarpine is placed on the back of a tick or the tiny varroa mite that parasitizes honeybees, they too will release their toxic secretions.

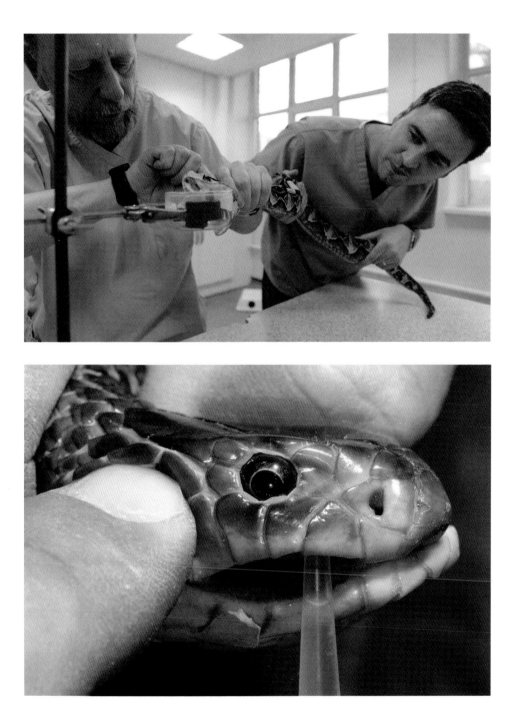

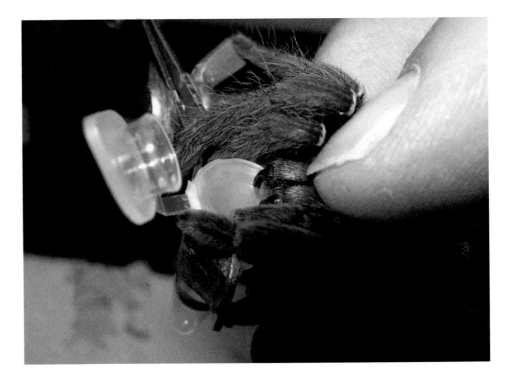

Milking venom from a blue fang tarantula, *Ephebopus cyanognathus*.
Although it can bite when sufficiently provoked, its first line of
defence is to flick barbed irritating hairs from its palps (the pair of legs
next to the chelicerae). This mode of defence is unique to this genus, as
other tarantulas flick hairs from their abdomen with their hind legs.

Getting venomous animals without fangs in the front of their mouths to secrete their venom does not always require chemical stimulants, however. Helodermatid lizards, such as the Gila monster and beaded lizards, as well as the large shrew-like solenodon, can be induced to salivate copiously when provoked to chew on a piece of rubber tubing. Researchers just have to collect the toxic drippings.

Similar techniques are used to extract venom from stinging animals. The exquisitely painful venoms of bullet ants and tarantula hawk wasps can be obtained by immobilizing them, and inducing them to sting into a parafilm-covered tube. Stonefish and toadfish can be milked by pushing down on the sheath of soft tissue that surrounds the venom spines on their backs, which leads to a powerful ejection of venom from the tips of the spines. To collect venom from stingrays and chimeras

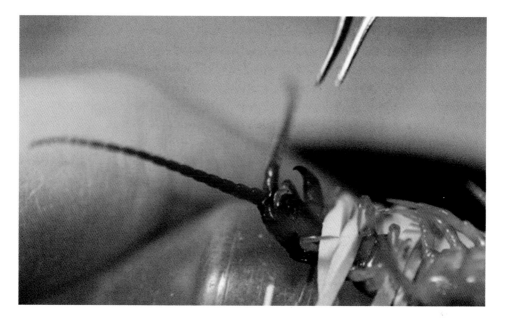

Milking venom from the flag-tailed centipede, *Alipes grandidieri,* by applying a mild electric current to the venom claws with an electric forceps. During the procedure the animal makes a hissing-like warning sound by vibrating its last pair of legs.

– cartilaginous fish also known as rat fish – researchers scrape the venom-producing tissue off the surface of the serrated spines.

Venom milking of most spiders, centipedes and scorpions, however, often involves electricity. If a mild electrical current is applied to the venom claws of centipedes, the jaws of spiders and the stinger of scorpions, venom is expelled as a result of the contraction of muscles surrounding the venom glands. Conveniently, bee venom can be collected by placing a glass-covered electric grid at the entrance of the hive. The electrical current induces the bees to sting when they land on it. Because electrical venom milking, when done carefully, does not harm the animals, they can be milked repeatedly to gather sufficient amounts of venom for scientific study. However, more invasive methods cannot always be avoided.

To get venom from a male platypus, for example, researchers insert a needle into the venom duct that runs along its hind leg and suck the venom into a syringe while applying gentle pressure on the venom glands to maximize the yield. Brave researchers wishing to collect wasp venom can immobilize them on a piece of sticky tape stuck to their finger, and then crush their heads with a pin to get them to release their venom. Some venoms, however, can only be harvested via dissection. To collect venom from bloodworms or aquatic cave-dwelling remipede crustaceans, their venom glands are dissected, and the venom is squeezed out manually using forceps. Similarly, after dissecting out the venom glands of wasps, they can be centrifuged to separate the venom from the tissue. In other cases, the small size or anatomy of the venom system forces the use of a cruder method of venom extraction. The small venom ducts of marine auger snails, jellyfish tentacles, or the needle-sharp bristles of *Lonomia* caterpillars, which are filled with devastating haemorrhagic venom, need to be homogenized to release their toxic ingredients. And to collect a large amount of venom from the fire ant *Solenopsis invicta*, researchers dig up a colony, and dunk the ants in a container with water that is overlain by a layer of an organic solvent such as hexane. As the ants sink through the solvent they eject their venom, which can then be extracted from the liquid.

Not all milked venoms are the same, not even when they are produced by the same animal. Different venom-milking techniques can result in different protein cocktails. For instance, compared to electrically milked venom, manually milked scorpion and honeybee venoms contain more contaminants as a result of squeezing the animals or dissecting their venom glands. Some animals can even secrete a different venom depending on whether it is intended for use in defence or predation, and different techniques may be needed to obtain each of them. To obtain the offensive venom from the deadly fish-hunting geography cone snail, *Conus geographus*, researchers bait it with a piece of fishtail attached to a tube covered with a condom. When the cone extends its proboscis to deliver its predatory sting it injects its venom forcefully through the piece of fishtail and into the receptacle. In contrast, to obtain the cone's defensive venom it is lifted out of the water and squeezed until it extends its proboscis to sting into another collecting tube. Cones can eject their venom with such force that you can hear it enter the tube. Remarkably, although a number of toxins are present in both the defensive and offensive venoms of the geography cone snail, each also contains a unique mix of toxins completely absent from the other. Unusually, the defensive toxin cocktail is the more complex of the two, as well as the most dangerous to humans due to an abundance of paralytic toxins How and why cones have evolved these different venoms is discussed in chapter 5, but before we can tackle such mysteries, we first need to have a closer look at how venoms actually work.

Predatory

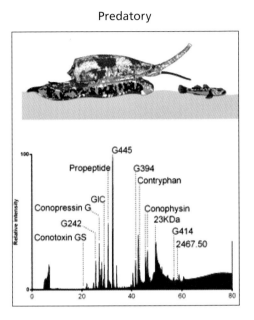

Defensive

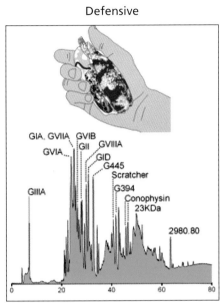

Composition

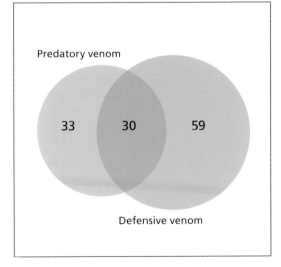

ABOVE LEFT A liquid chromatography-mass spectrometry (LC-MS) profile of the predatory venom of *Conus geographus*, with each peak in the graph denoting a venom component.

ABOVE RIGHT A LC-MS profile of the defensive venom of *C. geographus*.

LEFT The number of toxins in the snail's predatory and defensive venoms are indicated. The defensive venom is considerably more complex than the predatory venom, with 30 of the 63 predatory toxins also injected in defence.

Measuring the power of venom

During the last week of September 2015 the pubs of Oxford teemed with scientists with toxic interests. Oxford was hosting the 18th World Congress of the International Society on Toxinology, and the authors were there to present their latest research on centipede and bloodworm venoms, and to discover what their colleagues had been up to over the past two years. Anyone browsing the abstract book would quickly realize that venom researchers have eclectic as well as esoteric interests: 'Exposure of lactating rats to the *Tityus bahiensis* scorpion venom,'; 'Proteomic analysis of human blister fluids following enveriomation by three snake species,'; 'Tick toxins target vertebrate host wound healing.'

Although each of these studies looked at venoms with the myopic squint of the specialist, they had shared ambitions. They try to shed light on precisely how venoms work, how this benefits the venomous animal, and how it could benefit humans intent on harnessing the power of venom, such as in the development of new drugs. At the Oxford meeting, authorities on viper venoms rubbed shoulders with leech experts and tick specialists to explore how their animals wield their haemotoxic venoms to manipulate the flow of blood; coral snake aficionados found common ground with scorpion researchers to admire how the fast-acting neurotoxins of their favourite animals disrupt the functioning of nervous systems; and specialists on spider and ant venoms butted heads over which arthropod group would be most promising for the development of new, environmentally friendly bioinsecticides (see chapter 7).

Before we explore the mechanisms by which venoms target, manipulate, disrupt, damage and destroy the finely tuned physiological systems of organisms, we need to address a question that speaks strongly to the imagination. How lethal are venoms? The common currency for assessing the power of crude venom – venom that has not been fractionated into its constituent components – is known as the LD_{50} value, or the 'median lethal dose'. The LD_{50} is defined as the dose of venom needed to kill 50% of experimental animals to which it has been administered. It is measured as the amount of venom needed per unit weight of the envenomated animal, typically as milligrams of venom per kilogram of victim (mg/kg). The lower a toxin or venom's LD_{50} value, the more toxic it is. The commonest victims are lab mice. This makes sense because most LD_{50} experiments have been carried out on snake venoms and many snakes are predators of rodents. More importantly, however, we share many physiological traits with mice, and this allows a rough extrapolation of venom toxicity from mice to men, which is important for the development of effective antivenoms. A prospective antivenom that increases the LD_{50} of a venom in pre-clinical trials is promising for further development. In contrast, LD_{50} values for the venoms of spiders, scorpions and centipedes are often determined by injecting them into insects. This offers insights into how effective these venoms are at killing their intended prey, but it also gives researchers clues about the potential value of these venoms for the development of new insecticides.

Venom LD_{50} values are morbidly fascinating. It is therefore not surprising that they feature prominently on websites and in television programs. It is eerie how a small amount of venom can be lethal, and almost all the major animal groups in which venom has evolved are represented in the top toxicity rankings.

The deadly venom of the Mojave rattlesnake, *Crotalus scutulatus*, has an LD_{50} value of 0.2 mg/kg. Its primary toxin is a paralytic neurotoxin called Mojave toxin.

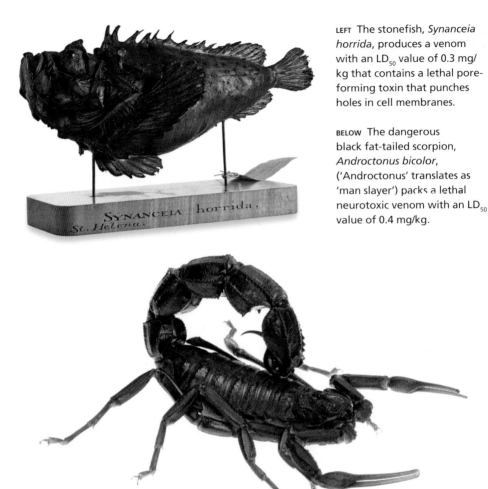

LEFT The stonefish, *Synanceia horrida*, produces a venom with an LD_{50} value of 0.3 mg/kg that contains a lethal pore-forming toxin that punches holes in cell membranes.

BELOW The dangerous black fat-tailed scorpion, *Androctonus bicolor*, ('Androctonus' translates as 'man slayer') packs a lethal neurotoxic venom with an LD_{50} value of 0.4 mg/kg.

If one compiled a complete ranking of LD_{50} values for all venoms that have been investigated – and most venoms haven't – many of the entries would belong to snakes, scorpions and spiders simply because their medically relevant venoms have been most intensely researched. But a complete ranking contains many surprises. For example, although insect venoms wouldn't rank particularly high, venom from the Maricopa harvester ant, *Pogonomyrmex maricopa*, which can be found in southwestern USA and Mexico, has an LD_{50} value of just 0.13 mg/kg when intravenously injected into mice. This makes it the most potent insect venom known, and ranks it well above most deadly snakes, including many vipers, mambas, cobras, kraits and sea snakes.

TOP The hylid frog, *Aparasphenodon brunoi*, has skin glands that contain potent toxins used for defence. The frog can envenomate enemies by jabbing and rubbing its head on them, which causes bony skull spines to puncture the toxic skin glands on its head, and deliver the toxins through the enemy's skin.

MIDDLE The skull of *A. brunoi* has bony spines near the nostrils, jaws and at the back (black arrowhead).

BOTTOM Magnified view of the bony spines near the frog's nostrils.

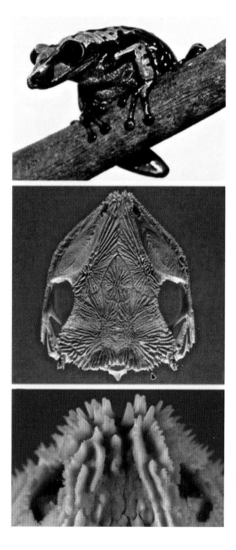

The innocuous looking casque-headed frog, *Aparasphenodon brunoi*, from Brazil packs a similar punch. This frog produces its venom in skin glands, and can deliver it by head butting. It has an unusually mobile head, and when it head butts an attacker, such as an intrepid herpetologist, bony spines on its skull penetrate the skin and inject the venom into the attacker, causing intense pain that can last for hours. When the frog's venom is injected into the abdominal cavities of mice, its LD_{50} value clocks in at 0.16 mg/kg, which is about 25 times more toxic than the average venom of deadly Brazilian pit vipers.

An LD_{50} value tells us the potency of a particular venom. But it does not reveal the killing power of the animal that produces it. The lethality of a venomous animal also depends on how much venom it can deliver in a single sting or bite. Its 'lethal capacity' can be defined as the amount of mice (or other animal) that would receive an LD_{50} dose of venom if all the venom in one individual were delivered in a single sting or bite. By this measure, some venomous animals are stunningly lethal. For example, a single Maricopa harvester ant can deliver an LD_{50} dose of venom to 200 g (7 oz) of mouse, or about 10 mice. The fearsome Japanese giant hornet, *Vespa mandarinia japonica*, has even greater killing power. Although gram for gram its venom is 40 times less lethal than that of the Maricopa harvester ant, the lethal capacity of a single hornet sting is 270 g (9½ oz) of mouse because it can discharge a much greater amount of venom than an ant sting.

However, ants and hornets rarely sting alone. To understand what could happen in the case of an all-out attack of a disturbed nest, entomologist Justin Schmidt has calculated the lethal capacity of entire hornet and ant colonies. A mature colony of one of the most feared of all hornets – the black-bellied hornet, *Vespa basalis* – houses an average of 5,000 workers. Their combined lethal capacity is 675 kg (1,490 lb) of mouse, or 33,750 mice. But this is dwarfed by the lethal strength of a colony of Maricopa harvester ants. With a conservative estimate of 10,000 workers, the lethal capacity of a large colony of ants would be 100,000 mice.

The giant hornet, *Vespa mandarinia*, is the largest hornet in the world. Giant hornets can deliver painful stings and they can eradicate entire bee colonies with their strong jaws, and feed them to their larvae.

The coastal taipan, *Oxyuranus scutellatus,* tops the lethality ranking of snakes. Its fast-acting venom attacks the nervous and circulatory systems. The venom causes the formation of lethal blood clots so quickly that bitten mice die before they even have a chance of becoming paralyzed.

Impressive though they are, the toxic power of these arthropods pales into insignificance compared to the damage that a single bite from an inland or coastal taipan could do. With venom that tops the LD_{50} ranking of snakes and a large venom yield, the intravenously delivered venom of a single coastal taipan, *Oxyuranus scutellatus*, bite has an almost unbelievable lethal capacity of 67,846 kg (149,574 lb) of mouse, or 3,392,307 mice. Not surprisingly, coastal taipans and their close relatives occupy the very pinnacle of venom virulence.

Most venoms, however, have probably not evolved to kill, although they may be very good at it. As long as defensive venom causes enough discomfort to deter a predator, and predatory venom disables prey long enough for it to be eaten, venoms don't have to kill. In addition, the effects of venoms are dose-dependent, and they can often achieve what they have evolved to do with a dose smaller than the lethal one.

The power of venom is context dependent

LD_{50} values provide fascinating insights into the raw power of venom, and are useful for antivenom development (see p.89), but they only tell a limited story. The power of venom is context dependent, and varies with site of injection, target organism and what the venom has evolved to achieve. The variable that determines the toxicity of a given venom injected into a given animal at a given time is the site of injection. The venom and envenomation behaviour of the Central American wandering spider, *Cupiennius salei*, have been particularly thoroughly researched. Researchers have discovered that its venom has different LD_{50} values depending on where it is injected into crickets. The closer it is injected to the central nervous system, the more lethal it is; the most lethal location for injection is the base of the first leg, while the top of the abdomen requires four times more venom to achieve the same killing power. The same is true for other venoms. The venom of the dangerous Brazilian wandering spider, *Phoneutria nigriventer*, has an LD_{50} value of 0.33 mg/kg when it is injected into a vein, but only 0.006 mg/kg when it is injected into the brain. Some snake venoms are more toxic to mice when injected into a vein, others when injected into the abdominal cavity. Some are more lethal when injected into muscles than when injected under the skin, and vice versa. Yet others kill more readily when injected under the skin than directly into a vein.

As well as lethality, the pain caused by envenomations also depends on the site of injection. In 2014 Michael Smith, then a graduate student at Cornell University, published a scientific paper that took the idea of curiosity-driven science to an unexplored extreme. The question that Smith felt needed answering was where it would be most or least painful to be stung by a honeybee. Over the course of several weeks Smith rated the pain of honeybee stings delivered to 25 locations on his body. He started each working day by allowing himself to be stung five times, and he scored the resulting pain on a scale between 1–10. After receiving hundreds of stings that covered the remotest corners of his anatomy – including the back of his knee, behind his ear, as well as his nipple, scrotum and armpit – Smith concluded that the worst locations to be stung are your nostril, upper lip and penis shaft. Smith scored the pain at these sting sites as 9.0, 8.7 and 7.3, respectively, which leaves room for other intrepid explorers to find entries for the top 10% of Smith's pain scale. In 2015, Smith was duly awarded, with Michael Smith, an Ig Nobel prize for this painful bout of self-experimentation. However, Smith was not the first biologist to take a somewhat masochistic approach to the study of insect venoms.

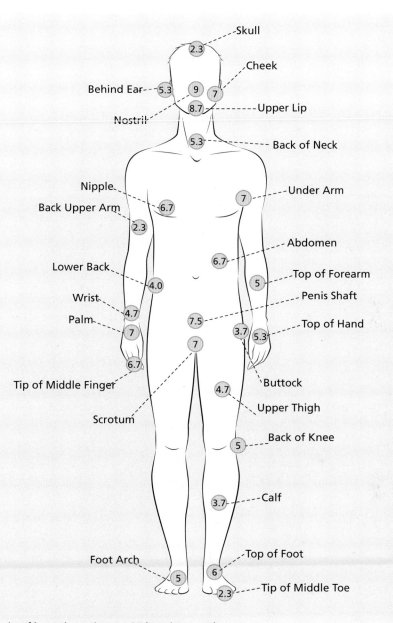

The pain of honeybee stings to 25 locations on the human body, on a scale of 1–10, as determined by Cornell University graduate student Michael Smith.

A PAINFUL BLUFF

Stinging insect expert Justin Schmidt relates many close encounters of the stinging kind in his wonderful book *The Sting of the Wild*. For instance, tarantula hawk wasps – large, conspicuous, spider-hunting wasps in the genera *Pepsis* and *Hemipepsis* – pack a sting that is even more painful than that of Maricopa harvester ants. Schmidt's 'tasting note' for their sting describes the pain as 'blinding, fierce and shockingly electric'. Schmidt devised a pain scale of 0–4 in an attempt to standardize the pain caused by hymenopteran stings.

In Schmidt's pain scale, tarantula hawk wasps are the only insects worthy of a score of 4 other than bullet ants and warrior wasps. You might therefore think that these wasps rank high in the LD_{50} chart. You'd be wrong. Despite their devastatingly painful stings, the LD_{50} value of their venom is unimpressive. More than 20 mg of venom from the most lethal species of tarantula hawk wasp is required to kill 1 kg (2 lb) of mouse. This makes honeybee venom seven times as deadly, and Maricopa harvester ant venom more than 200 times as lethal as the excruciatingly painful venom of tarantula hawk wasps. Schmidt therefore concluded that for mammals the lethality of tarantula hawk wasp venom 'is essentially nil'.

A tarantula hawk wasp tackles a tarantula. After delivering her paralyzing sting, the wasp will drag the spider to a burrow to lay an egg on it so that the wasp larva has a good supply of fresh meat to eat.

Although honeybee stings hurt much less than those of tarantula hawk wasps, they are much more dangerous because allergens in honeybee venom can cause allergic reactions, and in extreme cases fatal anaphylactic shock. The same is true for the European hornet, *Vespa crabro*. Despite their large size and warning coloration, like honeybees they only score a 2 on Schmidt's pain scale, but their venom contains the same potentially deadly allergens. Venoms that can trigger anaphylaxis have a disproportionate killing power because they augment the adverse effects of venom toxins with a catastrophic overreaction of the victim's immune system that can lead to respiratory failure and cardiovascular collapse. The anaphylaxis suffered by biologist George Madani after he was bitten by a slow loris, as related at the beginning of the chapter, may have been due to a protein in slow loris venom that is very similar to a known allergen present in cat saliva. While cat allergy manifests itself mostly as a runny nose and red eyes, delivery of slow

The European hornet, *Vespa crabro*, is the largest hornet in Europe, and is a predator of insects, including other wasps.

the animals are observed for 24 hours, but any deaths that occur hours after an envenomation event are unlikely to be relevant to a venomous animal in its natural habitat. Immediate pain or near instantaneous paralysis are useful traits in defensive and predatory venoms, and these powers have been exquisitely honed by natural selection. Their adaptive value is not diminished when pain eventually subsides and paralysis is reversed, as long as these effects last long enough for the venomous animal to escape its enemy or secure its meal.

Every year bites by venomous snakes kill around 100,000 people. But this doesn't mean that snake venoms have evolved to kill humans. The vast majority of envenomations kill their victims too slowly to benefit the snake. Indeed, snakes are often killed after they have bitten the victim, and being able to show the dead snake in hospital can be crucial in ensuring that the victim receives the correct antivenom. If defensive snakebites had evolved to kill large mammals like humans, which are often orders of magnitude larger than the snake's natural prey, snakebites should be able to kill a human quickly enough to minimize the chance of a fatal retaliation. This is rarely the case. Instead, snakebite victims typically succumb because of the continued effects of toxins that have evolved to rapidly subdue smaller prey. Unless these toxins are neutralized by antivenom, they continue to disrupt and destroy the victim's physiological systems.

A species of small jellyfish named *Carukia barnesi* can serve as an emblem of accidental human death in the world of venom. *Carukia barnesi* represents a group of about 25 species of jellyfish that can cause a unique constellation of envenomation symptoms, known as the Irukandji syndrome. When stung by these jellyfish, victims develop severe lower back and abdominal pain, nausea and vomiting, cramps and spasms affecting the whole body, difficulty breathing, profuse sweating, headaches, anxiety, and a sense of impending doom. Other symptoms may occur as well, and in fatal cases the victim dies of heart failure or cerebral haemorrhaging (stroke), which may occur several days after being stung. These effects on humans, however, in no way benefit the jellyfish. *Carukia barnesi* are rarely bigger than the size of a peanut, and although their four venomous tentacles can be 100 times longer than their bodies, even the largest specimens have no greater ambition than to catch and eat planktonic crustaceans and larval fish.

The small Irukandji jellyfish, *Carukia barnesi*, is a distant
relative of the deadly sea wasp, *Chironex fleckeri*. It changes
its venom composition as it matures, with immature individuals
catching invertebratres, and mature ones hunting fish. This one
is digesting two fishes.

The brown tree snake, *Boiga irregularis*, is a venomous snake
that belongs to the Family Colubridae. In contrast to snakes in
the families Viperidae and Elapidae, which are front-fanged,
the brown tree snake delivers venom with grooved fangs that
are further back in its mouth.

Natural selection hasn't maximized the power of venoms to kill humans, even though some venoms produced by animals as different as jellyfish, scorpions, spiders, snakes and molluscs are able to do just that. In fact, many venoms haven't evolved to maximize killing power at all. The ranking of LD_{50} values shows that most venoms don't even come close to the lethal potency of the very deadliest. The idea that natural selection hasn't shaped venoms purely to maximize killing power is supported by experiments that compare the LD_{50} values of crude venom and individual venom toxins. For example, crude venom of the brown tree snake, *Boiga irregularis*, has a substantially higher LD_{50} value (lower toxicity) than a purified sample of its major neurotoxin. The same is true for the venom of the black mamba, *Dendroaspis polylepis*. If the black mamba's venom consisted solely of its most potent toxin, it would be eight and a half times more lethal than it is. These venoms function well as weapons for the subjugation of prey, even though they could have conceivably been more deadly. If natural selection has not honed them purely for lethal power, what else have venoms evolved to achieve? As we will see in the next chapter, one answer is speed.

Speed of action is one of the outcomes of the evolutionary arms race between venomous animals and their victims. The ongoing evolutionary battle between envenomator and envenomated also provides the context needed to understand why venom varies in potency and composition depending on the sex, developmental stage and geographic location of the venomous animal. But before discussing this topic more fully in chapter 5, we first need to gain a deeper insight into the power of venom that goes beyond LD_{50} values.

Chapter 4

Dissecting the Power of Venom

V enoms are extraordinarily potent. Tiny spiders from the genus *Zodarion* pounce on ants that are more than 30 times heavier. The spider delivers a single quick bite to an ant's leg and then retreats to a safe distance to wait for the venom to take effect. The bite causes the ant to stand motionless with gaping mandibles as the bitten leg contracts. As the spider venom spreads through its body, the ant convulses and contorts into the shape of a comma, as its abdomen bends tightly underneath its thorax. After several minutes the ant is completely paralyzed and falls over, enticing the spider to approach again and start its meal. How can a tiny volume of venom be so powerful?

The answer is extreme target selectivity. To maximize the power of a weapon it is crucial to aim it at a vulnerable target. For example, a cheetah throttles an antelope with a throat bite; a secretary bird aims its deadly kicks at a snake's head. The same is true for venomous animals. To maximize the debilitating effect of its venom, a centipede prefers to bite an insect's head or thorax, and a spitting cobra will target an assailant's eyes. Indeed, as we saw in chapter 2, some animals, like the jewel wasp, *Ampulex compressa*, need to deliver their venom with surgical precision for it

The paralyzing venom of this lynx spider
allows it to capture prey substantially
larger than itself.

to be effective. But in the heat of the moment precision aiming is often impossible. A venomous animal under attack by a predator is unlikely to have time to choose where to land its teeth or stinger, and even agile predators may not be able to aim their strike with perfect precision at fast-moving prey. Luckily, most venoms work even if they are delivered to sub-optimal locations. The most important factor determining the target selectivity of venom is not how well venomous animals can aim their bites or stings, but the biochemical affinity that venom toxins have for specific targets in the body. The world of venom is a realm of self-guided missiles where even the blind are sharpshooters.

Targeting organs

Target selectivity can be considered on two levels: the higher level of entire organs and physiological systems; and the lower level of molecules. To understand the biological role of a venom – roughly, to answer *what* a venom has evolved to achieve, and *why* – it is instructive to consider its higher level targets. To appreciate *how* a venom manages to achieve its effects, you need to study it on the molecular level. Applying this strategy to predatory and defensive venoms, what is the quickest and most reliable way of deterring or debilitating an adversary or potential prey? The answer is to target the neuromuscular system – the nerves and muscles that control locomotion – or the cardiovascular system that controls blood flow. Why is this so? Because disrupting either system minimizes the chances of a victim being able to escape or retaliate.

Indeed, many venoms have evolved to attack these physiological Achilles' heels. Toxins that attack the neuromuscular system are generally referred to as neurotoxins, and disruptions of this system are referred to as neurotoxic effects. Common neurotoxic effects caused by venoms are rigid or spastic paralysis, caused by uncontrollable muscle contraction, or flaccid paralysis caused by blocking muscle contraction. Similarly, toxins that disrupt the flow or clotting of blood are known as haemotoxins, and their effects as haemotoxic effects. Other common labels used for toxins and their effects are myotoxic (damages muscles), cardiotoxic (affects the heart), cytotoxic (toxic to cells), necrotic (kills cells by rupturing cell membranes), haemorrhagic (causes bleeding), haemolytic (destroys red blood cells) and nephrotoxic (damages kidneys).

Although these labels describe the physiological effects of venom toxins, they do not demarcate neat and non-overlapping categories. For example, a haemolytic toxin that destroys red blood cells is cytotoxic and haemotoxic at the same time.

The venom cocktail of the Brazilian yellow scorpion, *Tityus serrulatus*, affects all physiological systems of envenomated humans. Among others symptoms and targets include convulsions (muscles), heart arrhythmias (heart), inflammation (immune system), salivation (exocrine glands), diarrhoea (digestive system), involuntary erection (reproductive system), hypertension (circulatory system) and pain (nervous system).

Some toxins fall into multiple categories. The peptide crotamine, for instance, is a neurotoxin present in many rattlesnake venoms that causes spastic paralysis of the hind limbs of rodent prey. But this neurotoxic effect, which it exerts through acting on nerves, is not the only damage it causes. It is also cytotoxic as it penetrates and kills cells, and it is myotoxic and necrotic as well, as it can cause severe muscle necrosis.

It is unwise to use labels uncritically to characterize entire venoms because venoms are complex toxin cocktails that can cause harmful effects across different physiological systems. To label cone snail venoms as neurotoxic is relatively unproblematic because, although they are very complex cocktails, the vast majority of venom components attack the neuromuscular system to cause paralysis. An extreme opposite of this neat neurotoxic scalpel is the venom of the sea wasp, *Chironex fleckeri*. This shockingly painful jellyfish venom delivers a seemingly indiscriminate battering ram assault on multiple fronts, and causes inflammation, neurotoxic, haemotoxic, cytotoxic, cardiotoxic, myotoxic and dermonecrotic (kills the cells of the skin) injuries. And to add insult to injury, sea wasp venom can kill a human faster than almost any other venom.

Molecular targets

The molecular interactions of venom components and their targets depend on such features as their electrical charge (positive and negative charges attract), the presence of hydrophilic and hydrophobic domains (parts of molecules that are attracted to and repulsed by water, respectively), and their 3-D molecular configurations. For venom proteins, these traits are determined by their constituent amino acids. For instance, arginine is a positively charged, hydrophilic amino acid that is often found on the surface of venom proteins where it is in contact with the watery environment. In contrast, leucine is a hydrophobic amino acid that is repelled by water, and is therefore typically found buried in the hydrophobic core of the folded venom protein. The clustering of such hydrophobic amino acids in a protein's core is important for stabilizing its folding, while amino acids exposed on the protein's surface are especially important for mediating interactions with other molecules. A protein's folding in space determines how and how strong it can interact with other molecules, and therefore determines how it functions.

A protein found in the Chinese scorpion, *Mesobuthus martensii*, illustrates these ideas. Scorpion haemolymph – the fluid that bathes the internal organs – contains a range of more or less related anti-microbial peptides collectively known as defensins. Defensins are a potent weapon of the immune system of many arthropods, including scorpions. One of the defensins of *M. martensii* is a potent killer of bacteria, but remarkably, it can also act as a neurotoxin by blocking ion channels. Ion channels are proteins wedged in the cell membrane. They enclose a central pore that can be opened or closed to allow ions (charged particles) to enter or exit the cell. The movement of ions, such as sodium, calcium and potassium ions, across the cell membrane of nerve cells drives the transmission of nerve impulses. By regulating this flow of ions, ion channels regulate the conduction of nerve impulses in the nervous system.

The scorpion defensin can block a type of potassium ion channel by binding to the region of the ion channel pore that faces the outside of the cell. This prevents potassium

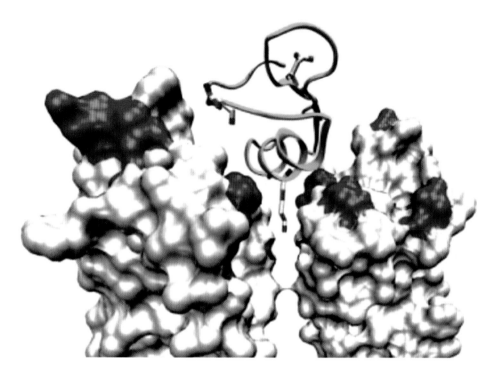

A 3-D model of the extracellular pore of a potassium ion channel.
This pore normally allows the movement of potassium ions across
the cell membrane via the ion channel, which is crucial for the
regulation of nerve impulses. When the flow of potassium ions
through the ion channel is blocked by a scorpion toxin (in green)
that blocks the pore, nerve impulses continue to be produced,
which can lead to paralysis by sustained muscle contraction.

ions from travelling through the channel out of the cell, with the result that the nerve
cell continues to generate nerve impulses. Neurotoxins such as this are common in
scorpion venoms, and cause the continued generation of nerve impulses that lead to
spastic or rigid paralysis of prey, characterized by uncontrollable muscle contractions.
This neurotoxic effect of the scorpion defensin depends crucially on just two of the
defensin's amino acids, which are critical for its binding to the ion channel. One of
them is an arginine residue that interacts with three amino acids of the extracellular
pore region of the ion channel, and the other is a lysine residue that actually blocks the
channel's pore. When these residues are experimentally changed to different amino
acids, the defensin's ability to bind to the ion channel drops by about 70%.

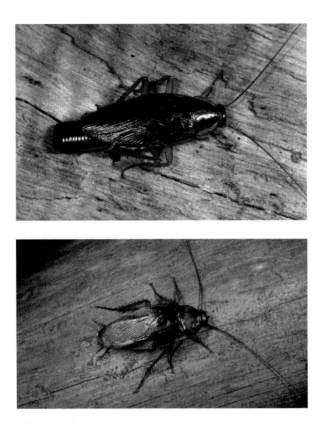

LEFT AND BELOW LEFT German and American cockroaches look similar, but they can have different sensitivities to spider venoms.

This example shows that the selectivity of the interactions of proteins with other molecules is finely balanced. Because each protein is comprised of a unique sequence of amino acids, with distinct properties, each protein has a unique fingerprint of molecular characteristics that determines whether and how it will interact with other molecules. Consequently, a tiny change in a toxin can radically change its target selectivity and physiological effects. Similarly, a small change to the molecular target of a toxin can protect it from attack. The venom of the African tarantula, *Augacephalus ezendami,* contains an insecticidal neurotoxin that can paralyze insects. Experiments with cockroach neurons show that it achieves this by blocking sodium ion channels, which prevents the transmission of nerve impulses to the insect's muscles. Although it has this effect on neurons of the German cockroach, *Blatella germanica,* neurons of the American cockroach, *Periplaneta americana*, are unaffected by the toxin. Remarkably, this difference in toxin sensitivity is caused by a difference in just a single amino acid in part of the ion channel of the American cockroach. Other insects, such as the triatome bug and Chagas disease vector, *Rhodnius prolixus,* are similarly immune to this toxin as a result of an identical mutation to their sodium ion channel.

On a molecular level venom toxins typically have narrow target selectivities, but these targets may be ubiquitous in a victim's body. The physiological effects of a toxin in a given situation therefore depend on the distribution of its molecular targets. The ability of toxins to home in on very specific targets with a body-wide function explains why even a tiny amount of locally delivered venom can wreak systemic havoc. After teeth, stingers, spines and claws have delivered their toxic cargo across the skin barrier, the venom encounters dense forests of nerves and sensory receptors and a rich supply of blood vessels. These provide immediate targets for attack, but blood vessels, assisted by the lymphatic system, also provide an effective transport system to carry the destructive power deep into the body.

While spreading through the body of their victim, neurotoxic peptides such as those dominating the venoms of cone snails and elapid snakes (such as cobras, kraits and mambas), home in on nerves and the junction between nerves and muscles. There they will activate or block ion channels and neurotransmitter receptors, dramatically increasing or reducing the transmission of nerve impulses, which can cause spastic or flaccid paralysis of the victim. This explains Joe Slowinski's tragic end as described at the beginning of chapter 3. When Slowinski was bitten by a juvenile multi-banded krait, he was injected with a tiny amount of very potent neurotoxic venom that blocked the transmission of nerve impulses to, as well their reception in, Slowinski's muscles. After the krait venom had circulated through his body for five hours, he could no longer breathe on his own because the muscles of his diaphragm were no longer contracting. Eventually, after his body had become fully paralyzed, his heart stopped as well. However gruesome, Slowinski's dying struggle was probably mostly painless, as was the snakebite itself.

There is a final sad irony. The reception of nerve impulses in skeletal muscles is mediated by a neurotransmitter receptor, known as the nicotinic acetylcholine receptor. One of the krait's most potent paralytic neurotoxins, called alpha-bungarotoxin, has such a strong affinity for this receptor that it was instrumental in isolating it and allowing scientists to understand the receptor's structure and function. Unfortunately, this became a fatal attraction for Joe Slowinski, with the toxin coursing through his body for hours until it had reached all its targets with deadly efficiency.

Blood vessels not only provide an efficient transport system for the spread of toxins, they are also targets themselves. Natriuretic peptides are vertebrate hormones that are potent regulators of blood pressure, and they act by binding to receptors in the walls of blood vessels, leading to relaxation of the smooth muscles in the blood vessel walls. They are also found in the venoms of a range of snakes, such as mambas and taipans, where they cause a rapid drop in blood pressure that assists in quickly immobilizing prey. This sheds light on Bryan Fry's early symptoms after he was bitten by a Stephen's banded snake, which has venom loaded with natriuretic peptides to trigger a debilitating drop in blood pressure in prey animals. Immediately following the bite, he experienced a pounding headache and lost consciousness. When he arrived in hospital his blood pressure had sunk to 78/26 which, if untreated, could have led to multiple organ failure. Interestingly, the toxic saliva of the vampire bat, *Desmodus rotundus,* also contains natriuretic peptides that play a role in causing the relaxation of blood vessels to facilitate blood feeding – the blood flows more easily from the site where the bat has pierced the animal it is feeding on.

Another class of fast-acting neurotoxins, known as sarafotoxins, have the opposite effect of natriuretic peptides. Sarafotoxins occur uniquely in the venoms of the enigmatic burrowing asps that we encountered in chapter 2, snakes in the genus *Atractaspis* that are sometimes also referred to as stiletto or side-stabbing snakes. These cardiotoxic peptides cause acute high blood pressure by the powerful contraction of coronary and other arteries.

Enzymes are proteins that convert a substrate into one or more products. Although enzymes usually do not bind to specific receptors in the manner of many neurotoxins, they are usually highly specific, and each enzyme only modifies particular molecules. Venom enzymes grab onto their preferred molecules wherever they can, and in breaking them down disrupt physiological processes, destroy cells, tissues and organs, and cause a plethora of symptoms. The enzyme hyaluronidase, for instance, breaks down hyaluronic acid and chondroitin sulphate, major components of the extracellular matrix that embeds the tissues and organs of vertebrates and most invertebrates. Hyaluronidase is present in the defensive and predatory venoms of a wide range of species, including centipedes, spiders, cephalopods, fish and reptiles. The damage done by this enzyme is indirect. By disrupting the integrity of tissues, it acts as a spreading factor that facilitates the penetration of toxins throughout the victim's body.

Venom enzymes with particularly devastating direct effects are the snake venom metalloproteases (SVMPs) that are especially prevalent in viper venoms. Although

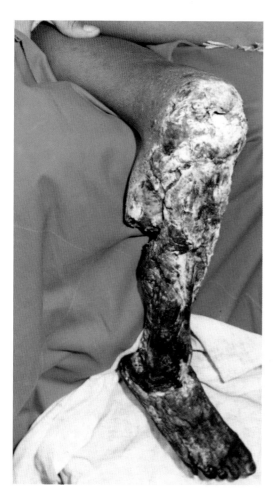

Extreme tissue necrosis in the leg of an 11-year old Ecuadorian boy two weeks after he was bitten by a *Bothrops asper* pit viper. He was treated with antibiotics, but this did not block the tissue destroying actions of the venom enzymes.

SVMPs can cause a variety of symptoms, most of them are strongly haemorrhagic. They break down components of the extracellular matrix that surrounds blood vessels, thereby destroying their structural integrity, and leading to internal bleeding that can quickly immobilize and kill prey. Because enzymes can remain active for long periods of time, venoms largely composed of enzymes, such as those of rattlesnakes and other vipers, can cause massive damage as shown above.

Returning to Bryan Fry's snakebite accident described in the previous chapter, when his extremely low blood pressure had been stabilized, he began to suffer more gruesome snakebite symptoms. First, blood trickled from his nostrils, then his eyes, and finally he started to bleed from his anus. These symptoms were caused by toxins with the exact opposite action of the blood-thinning toxins of leeches and mosquitoes. The blood-curdling venom of Stephen's banded snake contains a high concentration of enzymes that activate the blood-clotting pathway. In a small mammal this would rapidly cause a massive and deadly clot. But when diluted by a human's large blood volume, the snake venom's blood-clotting enzymes create an enormous number of tiny clots that spread through the victim. Because the venom uses up all clotting factors the victim is left with blood that cannot clot. One of the blood-clotting factors in Stephen's banded snake venom, called hopsarin D, converts the inactive precursor of thrombin into the active blood-clotting enzyme. Remarkably, it is a weaponized version of a normal blood coagulation factor (factor Xa) that was already present in the last common ancestor of snakes and mammals. Mammals have retained it in its original role, but Stephen's banded snake has recruited this enzyme to function as a toxin in its venom. As we will see in chapter 5, the modification and recruitment of normal proteins to function as venom toxins is a common evolutionary pathway.

Equally destructive are the non-enzymatic pore-forming toxins that are important components of the venoms of scorpaeniform fishes (including scorpion fish, lion fish and stonefish), cnidarians, assassin bugs, centipedes, some spiders and honeybees. As their name suggests, pore-forming toxins are proteins that causes the formation of pores in cell membranes. The pore-forming toxins of the stonefish *Synanceia horrida,* and the box jellyfish, *Chironex fleckeri*, for instance, wedge themselves into cell membranes, destroying the cells and causing a range of symptoms, including excruciating pain, inflammation, tissue necrosis and death. Alpha-latrotoxin, which is a pore-forming toxin present in the venom of black widow spiders, *Latrodectus* spp., causes severe pain by inserting itself into neuronal cell membranes where it acts as an ion channel. These ion channels allow an influx of calcium ions, triggering a massive neurotransmitter release by the nerve cells, and leading to generalized pain and muscle cramps.

Platypus venom contains toxins very similar to the pore-forming toxins of stonefish and black widow spiders. This may explain why the war veteran who was envenomated by a platypus, as related at the beginning of chapter 3, was becoming incoherent with pain an hour-and-a-half after being stung, a pain he reported being much worse than shrapnel wounds he had sustained during the war.

The major pain-causing toxin in honeybee venom is likewise a pore-forming toxin. It is a peptide called melittin that makes up about 50% of the dry weight of honeybee venom. By puncturing holes in cells, it triggers the release of pain-causing substances such as serotonin, but it can also directly activate the heat-sensitive pain

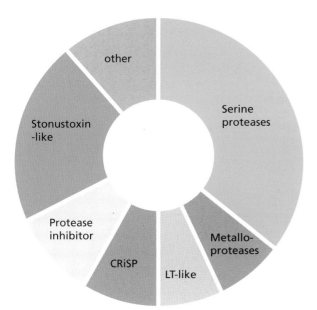

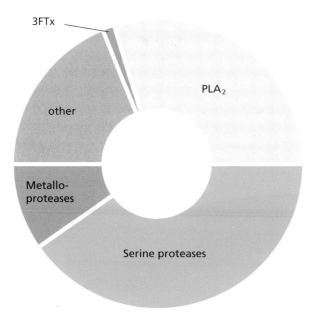

A comparison of the venom composition of the platypus, *Ornithorhynchus anatinus* (top), and a rattlesnake, *Sistrurus catenatus edwardsii* (bottom). The two most abundant types of components in platypus venom are enzymes (green and orange) and pore-forming toxins (blue and light grey). The rattlesnake venom is dominated by enzymes (green, orange and cream). PLA2 = phospholipase A2; 3FTx = three-finger neurotoxin; LT-like = latrotoxin-like protein; CriSP = cysteine-rich proteins.

BLOOD-SUCKING TOXINS

Despite the fact that envenomation by a mosquito isn't as dramatic as the playpus and honeybee, it involves an equally sophisticated toxin action. When mosquitoes are feeding, they inject a cocktail of anti-inflammatory and anticoagulant toxins to secure a good flow of blood, and to minimize the chances of detection by the host. The saliva of species in the genus *Anopheles*, many of which are malaria vectors, contains a peptide called anophelin, which is a potent inhibitor of blood clotting. When injected by a feeding mosquito, anophelin binds tightly to the thrombin enzyme that is prevalent in human blood. Thrombin is the final enzyme in our blood clotting pathway, and preventing its action blocks blood clotting. Anophelin binds both to thrombin's active site – the enzyme's business end that breaks down its substrate – as well as a patch of amino acids that functions as a recognition surface for thrombin's substrate. Other blood feeders, such as leeches, have evolved similarly effective anticoagulants. The medicinal leech, *Hirudo medicinalis*, for instance, secretes a peptide in its venomous saliva called hirudin that blocks both the active site and the same substrate recognition site as anophelin.

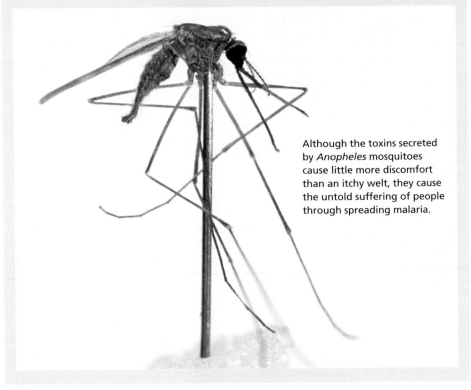

Although the toxins secreted by *Anopheles* mosquitoes cause little more discomfort than an itchy welt, they cause the untold suffering of people through spreading malaria.

receptor commonly known as the capsaicin receptor. This is an ion channel that can also be opened by the active ingredient of hot chilli peppers, which results in the sensation of heat.

The speed of venom

Most situations that require the deployment of venom require it urgently. When a small fish, known as the fangblenny, suddenly finds itself in the mouth of a larger fish it relies on its painful venomous bite to be spat out. Remipede crustaceans are blind and slow swimming predators of the pitch black waters of submarine caves. Their venom helps them to quickly overpower more agile crustacean prey, which is envenomated by a pair of powerful stabbing appendages. The toxic secretions of feeding mosquitoes and leeches as well as the peculiar blood-sucking vampire snail, *Colubraria reticulata*, must act swiftly enough to prevent their blood meals from curdling before they are fully ingested. And when a blue coral snake, *Calliophis bivirgatus*, comes face to face with a king cobra in a Malaysian rainforest, its exceptionally fast-acting venom becomes a matter of life or death, as these rivals are both highly venomous specialist snake eaters. Clearly, the need for speed has been an important factor in the evolution of venom.

Of course this doesn't mean that the enduring effects of venoms are devoid of adaptive value for the envenomator. Lingering pain may help to permanently imprint upon a predator's brain the message to steer clear of a particular venomous prey. The destructive power of enzymes in a predator's venom may also offer vital assistance to its digestive processes. And the ability of many hymenopteran and spider venoms to cause long-lasting immobility or complete paralysis of prey without killing it, is essential for providing a reliable meat larder for the animals or their larvae.

Irrespective of the importance of these persisting effects, the onset of which may be gradual and slow, the most important window of activity for virtually any venom is very narrow. It lasts just a couple of seconds to at most a few minutes. If an animal's defensive use of venom doesn't immediately deter an attacking predator, the animal risks injury or death. A predator whose venom fails to overwhelm its prey before it can escape will go hungry. This is especially important for slow or sessile predators hunting agile prey, such as sea anemones, cone snails, remipede crustaceans and arboreal snakes with a taste for birds. So how quickly do envenomation effects take hold?

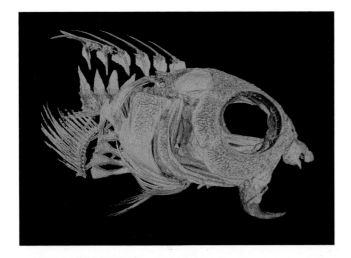

LEFT The fangblenny, *Meiacanthus grammistes*, is a small fish with a large punch. When grabbed by a larger fish it can deliver, with a pair of large venom fangs (coloured red), a defensive venom that lowers the blood pressure of the prey.

BELOW The blue coral snake, *Calliophis bivirgatus*, has enormously elongated venom glands that extend about a quarter of the length of its body. Its exceptionally fast acting venom allows it to prey on fast, venomous snakes.

Venoms that are used for defence, or as a weapon in fights between members of the same species for territory or mates, work by delivering near instantaneous pain. This pain may be ephemeral and bearable, like the sting of a sweat bee, or utterly debilitating and enduring, like a platypus sting, but the pain must start immediately. To achieve this, venom components act upon nerves and pain receptors, and do so in various ways. For instance, the pre-venom (the first droplet of venom that is ejected) of the African scorpion, *Parabuthus transvaalicus,* contains a high concentration of potassium ions, which induce pain by causing nerves to fire at the envenomation site. Hornet venoms contain components – such as the neurotransmitters serotonin and acetylcholine and peptides known as kinins – which all cause pain by directly activating pain receptors. The presence of serotonin likewise explains the instant sharp pain felt when wasps, as well as some spiders, cnidarians, centipedes and fish envenomate you.

Venomous animals have also evolved special toxins to target the pain pathway. The venom of honeybees, as well as that of several species of spiders, sea anemones and scorpions, contains peptides that cause immediate pain by activating the capsaicin pain receptor. Bites of the Texas coral snake, *Micrurus tener*, cause acute and intense pain. This is unusual for venoms in the group of snakes to which the Texas coral snake belongs (a group including cobras, mambas, taipans and kraits), which have evolved primarily as offensive weapons that cause rapid paralysis, often without producing pain at all. The venom of the Texas coral snake, however, contains a neurotoxin that potently activates acid-sensing ion channels that are present in the sensory neurons of skin and muscles. Activation of this acid-sensor causes intense and long-lasting pain in any mammal, bird, or reptile predator of the snake. Surprisingly, toxins recently discovered in the venoms of black and green mambas block these acid-sensing receptors, offering promising new avenues for the development of new painkillers.

Predatory venoms face a greater challenge. Local pain, however intense, is unlikely to be powerful enough to disable prey. But as we have seen earlier, predatory venoms can rapidly disrupt physiological functioning by targeting the two systems that are most immediately responsible for keeping bodies up and running: the circulatory and nervous systems. Australian elapids, such as taipans, *Oxyuranus* spp. and brown snakes, *Pseudonaja* spp., have venoms with powerful blood-clotting toxins that can cause rapid formation of blood clots and complete depletion of clotting factors, with prey succumbing to stroke injuries that result from clots blocking blood vessels. These coagulation-activating toxins are essentially turbo-charged versions of the clotting factor present in the venom of the Stephen's banded snake that we encountered earlier. In addition to possessing a weaponized version of coagulation factor Xa, taipan and brown snake venom also contains clotting factor Va, which acts as a cofactor to greatly increase the potency of clotting factor Xa. Moreover, as Bryan Fry experienced when a Stephen's banded snake bit him, a rapid drop in blood pressure can also be a very effective means of immobilization.

But in terms of pure speed, toxins acting on the nervous system are generally unbeatable. Readers who have seen film footage of snakes, spiders, scorpions and cone snails envenomating their prey will have witnessed the awesome speed with which neurotoxins can immobilize prey. The fastest of them take at most a few seconds to disable prey. The aptly named assassin bugs provide some impressive examples. The large African red spotted assassin bug, *Platymeris rhadamanthus*, is popular with amateur insect keepers. It is an insect predator that can rapidly paralyze a great diversity of insects. After being bitten by an assassin bug, American cockroaches cease their struggle after just 3–5 seconds, while larger prey, such as caterpillars and millipedes, are paralyzed within half a minute.

The nymph of the assassin bug *Rhynocoris carmelita* can paralyze the larva of the Mediterranean flower moth, *Ephestia kuehniella*, within 10 seconds. This is a remarkable feat given that the moth larva is 400 times heavier than the assassin bug nymph. Some assassin bugs can dispatch vertebrates just as quickly. The Middle Eastern assassin bug, *Holotrichius innesi*, for instance, is an insect predator, but its venom turns out to be twice as lethal to mice than it is to insects. This bug can cause respiratory paralysis and death in a mouse in 15–30 seconds with a single envenomation. Its venom is even deadly when inhaled, killing a guinea pig in two minutes.

But assassin bugs aren't the only predators with high velocity venom. The bite of long-legged house centipedes (Family Scutigeromorpha) immediately immobilizes insect prey. The aquatic maggots of horseflies deliver venomous bites that can paralyze insects within one or two spasms. Cuttlefish and octopus venom causes the onset of convulsions in crustacean prey within seconds, and crotamine, a neurotoxin that is present in the venoms of many rattlesnakes, takes only seconds to cause spastic paralysis of rodent hind limbs.

The fastest venoms have been forged in the fires of the extreme selection pressures involved in the arms race between predators and their agile or dangerous prey. Predators such as fish-hunting cone snails, bird-hunting mambas and the snake-hunting blue coral snake, *Calliophis bivirgatus*, have venoms that strike with the speed of a nerve impulse by activating sodium ion channels in nerve cells. This

The house centipede, *Scutigera coleoptrata*, is a fast moving predator of insects, which it paralyzes with its neurotoxic venom. This one is dining on a crane fly.

causes immediate, prolonged and uncontrollable muscle contractions that result in spastic or rigid paralysis. A fish that is harpooned by a cone snail has only a second or two before its fins are paralyzed, followed by full body paralysis just seconds later.

Taipan venom presents an intriguing departure from the expectation that neurotoxins will more quickly subdue prey than haemotoxins. It contains taipoxin, a presynaptic neurotoxin that inhibits the transmission of nerve impulses to muscles, causing flaccid paralysis. It is one of the most potent neurotoxins known in snake venoms, with an LD_{50} of just 0.002 mg/kg. In terms of sheer mouse killing power, it is six to 10 times stronger than crude taipan venom, and it can comprise more than half of all protein present in the venom. However, when mice are injected with an identical quantity of either pure taipoxin or crude venom, the animals receiving the crude venom die significantly more quickly. Moreover, when mice are injected with a dose of venom that corresponds to the quantity of venom that can be expected to be present in a natural taipan bite, the animals collapse and die almost immediately. They do not show signs of paralysis, however. Only when mice are injected with a low dose of venom do they become paralyzed before they die. Autopsies of the mice injected with venom reveals that their lung tissue is riddled with blood clots. This suggests that the taipan venom's powerful clotting factors killed the mice by obstructing the blood flow to their vital organs before paralysis could occur.

Why would taipan venom have this killing redundancy? If prey is despatched with powerful clotting toxins, does the paralytic taipoxin perhaps have another function? One plausible speculation is that taipoxin assists in the digestion of prey. Taipoxin is myotoxic and it destroys the integrity of muscle cells. It achieves this with enzymatic activity. It is composed of three sub-units, which are phospholipase A_2 enzymes that can destroy cell membranes by breaking down their phospholipids. Since the taipans' predominantly mammalian prey has substantial muscle mass, taipoxin may facilitate its digestion.

Predators like the ones discussed in this section are able to kill within minutes. But speed can come at the expense of endurance. Remarkably, several venoms have evolved toxins that work together to produce effects that are both lightning fast and enduring.

The Vietnamese centipede, *Scolopendra subspinipes,* produces a complex venom that is strongly insecticidal. Its bites can cause intense pain and swelling in humans that can last for several days.

The synergism of venom toxins

Just as the magic of a good curry is created by the combination and interplay of its ingredients, so does a syndrome of envenomation symptoms represent a toxic symphony. Venom toxins work together in various ways, complementing, enabling and enhancing each other's actions. Toxin synergism happens when toxins interact directly or indirectly to produce an envenomation syndrome that is more than the sum of the effects of the individual toxins. Toxin synergism is best understood in snakes, spiders, scorpions and cone snails, but it is likely a feature of complex venoms generally. The power of synergism can be enormous.

The hunting spider, *Cupiennius salei*, is probably one of the best studied spiders in the world. Researchers continue to study nearly all aspects of its biology, including its venom, behaviour, embryology, and the biomechanical properties of its venom fangs.

A Sydney funnel-web spider, *Atrax robustus*, adopts its threat
pose, with its front end raised and venom droplets emerging
from its chelicerae.

The crude venom of the Chinese red-headed centipede, *Scolopendra subspinipes mutilans*, is 3,500–50,000 times more insecticidal that its isolated neurotoxins, and synergistic interactions between toxins may be responsible for much of this difference.

A simple form of toxin synergism is illustrated by spreading factors, such as the enzyme hyaluronidase. Spreading factors facilitate the spreading of venom toxins through a victim's body, typically by disrupting the integrity of connective tissue around the envenomation site. When the main neurotoxin from the hunting spider *Cupiennius salei* is injected into insects together with the spider's venom hyaluronidase, it is more than twice as lethal than if it were injected alone. By destroying the integrity of connective tissue, hyaluronidase helps the neurotoxin reach its molecular targets. Hyaluronidase has been shown to play a similar potentiating role in snake venoms. In order for it to be able to kill mice, crotoxin – the major neurotoxin in the venom of the South American rattlesnake, *Crotalus durrissus terrificus* – requires hyaluronidase to pave the way to its targets.

The venom of the spitting African rinkhals cobra, *Hemachatus haemachatus*, provides another example of toxin synergism, but this time to target the blood-clotting cascade. The venom of the rinkhals contains two proteins, hemextin A and B, which are harmless on their own. Hemextin A is a mild anticoagulant and hemextin B is inactive. However, when two molecules of each subunit combine, a four-part toxin complex, called hemextin AB, is formed, which is a potent inhibitor of blood clotting. Finally, taipoxin from taipan venom, which we discussed earlier, is also an example of synergism. Only one of its three subunits is strongly toxic, but only when all three come together does it achieve a toxicity towards rodents unmatched by virtually any snake venom toxin.

Toxin synergism also allows venomous predators, such as snakes, spiders and cone snails, to specialize in capturing agile prey in environments where they can easily escape. Their venoms pack the toxic equivalent of the classic 'one-two' boxing combination. A quick jab followed by a hook is the perfect marriage of speed with knock-out power. Australian funnel-web spiders in the genera *Atrax* and *Hadronyche*, for example, have two neurotoxins in their venoms that complement each other in terms of how quick they act and how long their toxic effects last. The fast-acting neurotoxin causes immediate paralysis, but its effect disappears in several hours. The other neurotoxin takes 20 to 30 minutes to take effect, but its paralyzing power is irreversible.

Mambas employ a similar strategy. They often hunt birds in trees, and first throw a fast punch with neurotoxins that cause the immediate uncontrollable contraction

The cone snail, *Conus purpurascens*, practices the so-called 'taser-and-tether' strategy of hunting fish. Its very fast acting venom causes near instantaneous paralysis in which the fish's muscles contract (note the stiff fins on the prey fish), allowing the snail to swallow it.

of muscles, followed by a slower building uppercut of toxins that progressively block the transmission of nerve impulses to the muscles. This transforms a rapidly immobilizing cramp into a permanent flaccid paralysis.

The athletic abilities of the sluggish cone snails are even more mismatched with that of their agile fish prey. Remarkably, cone snail venoms also rely on toxin synergism to provide them with the 'one-two' combination of speed and staying power. Species such as the magician's cone, *Conus magus*, employ a regiment of toxins, known as the 'lightning cabal', which causes immediate and massive muscle cramping as if the fish had been tasered. They deliver this toxic cocktail by shooting the fish with a hollow harpoon that is tethered to the snail. The harpoon is a modified tooth of the radula, which is a tongue-like feeding organ. The resulting rigid paralysis then gradually becomes flaccid as another battalion of toxins, known as the 'motor cabal', spreads through the victim's body, terminating the transmission of nerve impulses and relaxing the fish's muscles.

Two species of fish-hunting cones, the deadly geography cone and its close relative the tulip cone, *C. tulipa*, employ a very different, but equally effective 'one-two' strategy. As they approach a fish these snails extend their rostrum, which is a tube-like structure in front of the mouth that can be distended hugely, and waft a collection of toxins into the water, known as the 'nirvana cabal'. These peptides are thought to cause a state of sensory deprivation and hypoglycaemic shock, and they rapidly cause fish to become inactive, which allows the snail to approach and engulf its prey. Once the prey has been engulfed by the snail's rostrum, the snail follows up with an injection of paralyzing venom. Unlike the harpoon of the magician's cone, the harpoons of the geography and tulip cones are not tethered to the snail, so they function more as arrows that the snail fires within its own mouth. The secreted cloud of toxins even allows a single snail to subdue and engulf multiple fish at a time, which the snail can then harpoon individually. The recently discovered identity of one of the components of the nirvana cabal is truly surprising, and how it has evolved is extraordinary.

Like other animals, cone snails use the hormone insulin as a signalling molecule involved in energy metabolism. This insulin is expressed in the central nervous system of the snails and its structure is very similar between species. But in addition to this normal insulin, the geography and tulip cones secrete a different insulin molecule in the nirvana cabal of their venom. This particular type of insulin is smaller than the normal mollusc insulin, and its structure and function mimic fish insulin. When it is absorbed through a fish's gills it induces sensory deprivation and hypoglycaemic shock, so that the fish become inactive, which allows the snail to engulf the stunned fish before stinging it. A comparison of insulins across a wide range of species strongly suggests that early in the evolution of cone snails the normal insulin gene was duplicated, and that one of the copies was recruited as a toxin into the venoms of the geography and tulip cones. Moreover, the origin of these venom insulins was associated with a marked increase in their rate of evolution, as they became increasingly dissimilar from normal mollusc insulin and started to resemble fish insulin.

The duplication of genes, the acceleration of evolutionary rates, and changes in molecular structure to match a new target – in this case the insulin receptor in fish – turn out to be general features of toxin evolution. In this chapter we encountered a snail hormone and a snake blood-clotting factor that were weaponized into venom toxins. The next chapter will explore more deeply how evolution fashions bog-standard proteins into the most potent of toxins.

Chapter 5

Evolving Venoms

Venom was among the first weapons used in the arena of animal evolution. It beat stealth, strength, speed and cunning by millions of years, and it allowed animals stuck on the seabed or living as fragile gelatinous floaters to colonize the niche of first predators. Sea anemones, jellyfish and their cnidarian kin have used venom for well over half a billion years to procure their food, defend themselves and compete with neighbours. But as this book has shown, venom is such a useful adaptation that cnidarians weren't alone for long in wielding its toxic power.

Life is extremely adept at independently finding similar ways to overcome related life challenges, and to make use of comparable opportunities. Even complex evolutionary marvels such as image-forming eyes or wings have emerged multiple times in animals on different branches of the tree of life. The independent evolution of similar traits in different taxa is called convergent evolution, and the world of venom presents a treasure trove of examples. It is perhaps not so surprising that similar evolutionary solutions have been reached given the set of strict rules that are laid out by Mother Nature. For instance, gravity demands certain design features to be shared by wings, whether they are an insect's or a

Social wasps tending and guarding their most precious commodity: a nest with developing young.

bird's, and the properties of light constrain the design of functional eyes. When you also consider that the process by which these solutions are arrived at – evolution – is fundamentally the same across species, and that many species have a similar genetic toolkit, the convergences of animal survival strategies and adaptive traits become much less surprising. That is not to say that convergent evolution is any less fascinating. The apparent ingenuity of convergent evolution is truly one of Nature's wonders. It is also key to our understanding of how venoms have evolved and continue to do so.

A handful of bricks to build a thousand buildings

In introductory biology courses taught at schools and universities, wings and image-forming eyes are often used as examples of traits that have independently evolved in different groups, such as insects and birds. However, few traits show such a degree of convergence as venoms and toxins, both in terms of the number of times they have evolved, and in terms of the similarities between them. The number of times animals have evolved ways to enter the chemical battlefield through the use of venoms is astonishing. Venoms have not only evolved in about one-quarter of all animal phyla but, within most of these, venom has evolved independently on several occasions. Just among ray-finned fish (Class Actinopterygii), for example, which include the less than pretty marbled stargazer (see p.42), venom has evolved on 11 separate occasions from different non-venomous ancestors. However, this number is dwarfed compared to arthropods, where venom has evolved independently over 40 times.

Given the number of times venom has evolved, it is hardly surprising that there is a high degree of convergence. Take the delivery mechanisms, for example. Venom is generally delivered through the creation of a wound, often via a bite or a stab wound. Within each group of venomous animal there are only so many body-parts that can be modified into structures capable of delivering venom in this way. Moreover, venom-delivery mechanisms are often related to the ecological roles venoms play. For instance, venoms primarily used for prey capture and/or feeding are usually delivered by structures such as teeth, which are already functionally involved in these tasks. Purely defensive venoms are usually delivered through modified physical defence structures, such as spines. When venom-delivery mechanisms are looked at in this way, it is only natural that there are similarities.

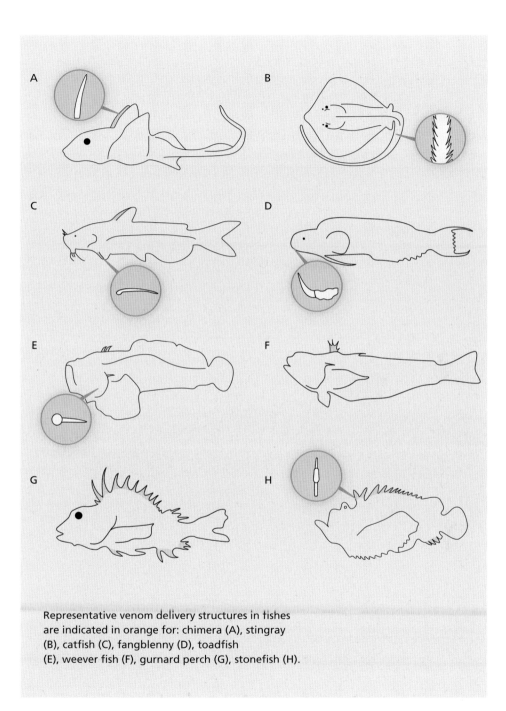

Representative venom delivery structures in fishes are indicated in orange for: chimera (A), stingray (B), catfish (C), fangblenny (D), toadfish (E), weever fish (F), gurnard perch (G), stonefish (H).

Among the ray-finned fish, which use their venoms almost exclusively for defence, eight of the 11 groups that have independently evolved venom have done so by modifying fin rays into venom-bearing spines. In addition to modifying their fin rays, both the Amazonian toad fish, *Thalassophryne amazonica*, and the mainly European weever fishes (Family Trachinidae), such as the ominously named *Echiichtys vipera* ('viper-like viperfish'; the lesser weever fish), have also modified their gill covers so that each one bears a venomous spine. In the case of *T. amazonica*, the gill cover and fin ray spines are remarkably similar. Both are surrounded by venom-producing tissue that feeds into the base of the hypodermic needle-like spine, a fascinating case of convergent evolution within a single venomous organism.

The convergence in the shape of the two sets of spines in the Amazonian toad fish is by no means unique. Injection of venom is in most cases the most effective method of delivery, and animals have evolved several ways of doing this. From venom-bearing

Envenomations by the dorsal fin or gill cover spines of the lesser weever fish, which occurs all along the British coast, cause intense pain that can last from an hour to three days. The most effective treatment is immersing the afflicted limb in as hot water as the patient can tolerate for up to half an hour.

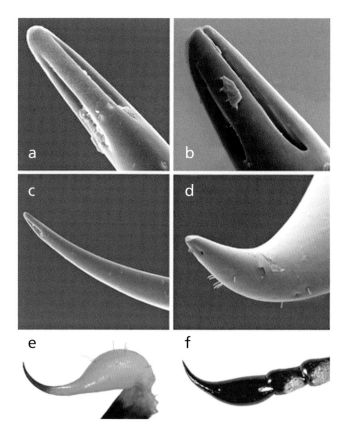

The longhorn beetle, *Onychocerus albitarsis,* has evolved an antennal stinger (b, d and f) to deliver defensive venom that is astonishingly similar to a scorpion stinger (a, c and e). The tips of the stingers are shown in (c) and (d) and under higher magnification in (a) and (b).

spines to highly specialized structures such as those of cone snails, most resemble hypodermic needles in some way or another, even though they may have completely different evolutionary origins (see chapter 2). The most remarkable example of such convergence is perhaps the similarities between the venom delivery apparatus of the longhorn beetle, *Onychocerus albitarsis,* and scorpions. Like other members of its genus, *O. albitarsis* is equipped with sharp spines on the tips of its antennae. However, unlike other known *Onychocerus* spp., the spines have been modified into venom delivering stingers that are surprisingly similar to the stingers of scorpions. Close examination renders them almost impossible to distinguish. Not only are the longhorn beetle's stingers a similar shape to those of scorpions, but they even have the same set of paired ducts opening through pores just short of the tip of the sting. Take a step back, however, and the differences are clear: one is found at the tip of the antennae of a beetle and the other at the tail end of an arachnid.

Building a chemical arsenal

The rampant convergence of venom systems is not just a biologically fascinating phenomenon; it also provides important insights into the evolutionary processes and forces at work on a molecular level. Just as venom-delivery systems can be identified as modified body-parts, the evolutionary origins of toxins can be traced back to proteins that perform everyday physiological roles, such as the regulation of blood pressure and blood clotting, the metabolism of proteins, and the defence against microorganisms. Sometimes referred to as 'housekeeping' proteins, they perform such important functions that they have remained largely unchanged over eons of evolutionary time, and they can usually be recognized even between very distantly related species. However, when a housekeeping protein is recruited to a venomous function, it can start to change, and often diversifies further to help the venom achieve its ecological role. Indeed, much of what we know about the ancestry of toxins comes from knowledge about our own bodies and a few other well-studied species. Even humans have proteins and peptides that could potentially be 'weaponized' to become part of a venomous biochemical arsenal, if it were advantageous for evolution to take us in that direction.

The bodies of even distantly related animal groups share a general molecular make-up that they have inherited from their common ancestor. Because of this evolutionary conservation, the recruitment of new toxins is limited by the number of suitable starting molecules. Many venoms also fulfil very similar ecological roles, and do so by containing toxins that target the same receptors and pathways. This limited number of both evolutionary starting molecules and functional end-properties means that, just like venom delivery mechanisms, toxins show a great degree of convergence both in terms of origin and of function.

Examining the proteins and peptides that have been weaponized into toxins can tell us a great deal about the properties that make a good toxin. Not surprisingly, these properties include high solubility (so they can be delivered in a venom solution), stability (so they retain the functional folding of the molecule) and involvement in some key, relatively fast-acting physiological process (so they can achieve rapid physiological disruption). Phospholipase type A_2 (PLA_2) proteins, for example, are highly stable, fast-acting enzymes that degrade phospholipids of cell membranes and release inflammatory and pain-inducing lysophospholipids and fatty acids. They have been recruited to venoms on at least 12 occasions in different parts of the animal tree of life. In contrast, keratin, a key structural protein in skin, is not likely

THE WEAPONIZATION AND EVOLUTION OF TOXIN GENES

Recruitment of most housekeeping proteins to a venom is thought to be a process that involves turning on the expression of an ancestral housekeeping gene in the venom-producing tissue, followed by gene duplication to yield multiple copies of the gene. Retaining one unchanged gene copy alleviates the natural selection pressure that acts to maintain the function and integrity of the ancestral gene, which is still required elsewhere in the body. The second copy of the gene that is expressed only, or mostly, in the venom gland is then free to accumulate changes so it can evolve new functions as a toxin, a process that can be called weaponization. As toxin gene families grow through further rounds of gene duplication, the relaxed selection pressure to maintain the original function not only means that some gene copies acquire new functions, but also that other copies accumulate mutations that make them non-functional. Toxin genes are, therefore, thought to evolve by a 'birth-and-death' model: new copies with new functions are born, while old copies that lose their function or ability to express 'die'.

to become a venom toxin as it lacks all of the above requirements. As it turns out, the proteins and peptides that fulfil all of the criteria required to be effective toxins appear in just a few protein and peptide classes. The resulting degree of convergence in terms of proteins and peptides recruited as venom toxins is quite stunning: all known animal venom toxins, as well as the tens of millions of toxins that must exist, can be assigned to just over 70 types of proteins and peptides!

Among the proteins that have made their way into venoms, many are enzymes that break down molecules such as proteins (proteases), lipids (lipases) or carbohydrates (glycoside hydrolases). Because they often perform the same action as their housekeeping counterparts, enzyme toxins tend not to be heavily modified, and may even be encoded by the same gene as the housekeeping version. Hyaluronidase is one such example. As mentioned in the previous chapter, hyluronidases break down hyaluronic acid, which is one of the main components of the gel-like extracellular matrix (ECM) that fills the space between our cells and organs. Hyaluronidases are involved in several important processes that range from embryonic development to wound healing. Most vertebrates have several genes that encode various hyaluronidases and in snakes such as the king cobra, one of these forms has also taken on a venomous role. Its substrate, hyaluronic acid, is usually found in the form of enormous molecular chains that create a viscous gel-like substance. When hyaluronidase breaks down these long chains, the ECM loses viscosity, which allows other toxins in the venom to spread more easily. By breaking down the ECM of the skin and other tissues, venom hyaluronidase also contributes to tissue damage and cell death in the area that surrounds the bite site.

Although enzymes that have been recruited to venoms have generally not changed much over evolutionary time in order to maintain their activity, they have often undergone subtle changes that have altered what tissues or substrates they target. For example, some of the main protease toxins found in the nematocysts of the starlet sea anemone, *Nematostella vectensis*, have been modified from proteases that are crucial to a number of developmental processes, such as controlling the development of symmetry. These proteases perform similar vital functions in organisms throughout the animal kingdom, and so have remained similar between groups of animals. The proteases found in the nematocysts of the starlet sea anemone, however, have lost sections (so-called 'domains') that control target specificity and ensure that only certain substrates are broken down. Losing these domains has allowed the venom proteins of the starlet sea anemone to become effective toxins, free of the molecular leash that once tightly controlled the actions of the housekeeping ancestral protein.

Not all enzymes conserve their ancestral activity while evolving into molecular killers, however. Some snake venom PLA_2 enzymes, for example, have lost their enzymatic activity, but they can still exert their toxic roles. PLA_2 toxins are among the main toxins of many snake venoms, and can act as myotoxins (destroying muscle cells), haemotoxins (destroying red blood cells) and even neurotoxins (disrupting the nervous system). Although many of these toxic functions rely on enzymatic activity, one of the two main myotoxic forms of PLA_2 found in viper venoms is enzymatically inactive. But despite having lost its enzymatic activity through a mutation of one amino acid, this PLA_2 can still damage muscles via a different mechanism. The ancestral enzyme provided an 'evolutionary scaffold' that allowed its toxic action to evolve while maintaining its stability and solubility.

Small but deadly

The concept of evolutionary scaffolds is particularly important for understanding the evolution of peptide toxins, which are the smaller proteins that dominate venoms such as those of spiders, scorpions and cone snails. As mentioned previously, stability and solubility are two key requirements for toxins. Since toxins are almost always recruited from secreted proteins, their solubility depends in large part on their stability. In order to remain stably folded, most proteins have a core of water-insoluble amino acids, and most changes to this hydrophobic core make the protein less stable or soluble. However, proteins rich in the amino acid cysteine are not so constrained. Cysteines contain sulphur atoms that readily make molecular bonds with sulphur atoms of other cysteines. These bonds are called disulfide bonds and provide highly stable, internal cross-braces that help stabilise the hydrophic core of the vast majority of toxins in animal venoms. In peptides, which are by definition smaller than proteins, that contain a large proportion of cysteines ('cysteine-rich peptides'), these disulfide bonds often account for most if not all of the hydrophobic core, and thereby also the structural integrity.

Due to their importance in maintaining protein structure, the cysteines in peptides are usually extremely well conserved over evolutionary time. The disulfide bonds essentially provide the structural scaffold on which evolution can take place. Because so much of the 3-D structure of these peptides depends on their disulfide bonds, most of the remaining amino acids can be mutated without damaging the overall structure. In other words, cysteine-rich peptides are extremely 'evolvable', and ideal candidates for weaponization.

This evolvability has a huge impact on the evolution of venoms, especially in animals that rely largely on peptide toxins such as spiders, scorpions and cone snails. Their venoms are normally extremely diverse, sometimes containing over a thousand unique molecules. However, because peptide toxins are so modifiable, little trace of their ancestral amino acid sequence may be retained. Fortunately, the conservation of structural scaffolds means the evolutionary origin of the peptide toxins can be determined using their 3-D structures. This has revealed that most cysteine-rich peptide toxins originated as molecules that were related to the immune system (collectively referred to as 'defensins') or as signalling molecules, such as neuropeptides or hormones. It is easy to see how such molecules could become useful toxins. Anti-microbial defensins, for example, act by destroying the membranes of intruding microbes, or by targeting their ion channels, either of which can be useful activities of a venom. Indeed, the main peptide toxin families of several venomous animals, such as scorpions and spiders, have evolved from defensin molecules.

Although both spiders and scorpions have recruited defensins as toxins, a comparison of their 3-D structures makes it clear that the toxins have different structural scaffolds, and hence originate from completely unrelated molecules. In spiders, they fold into a form that is called the inhibitory cystine-knot (ICK),

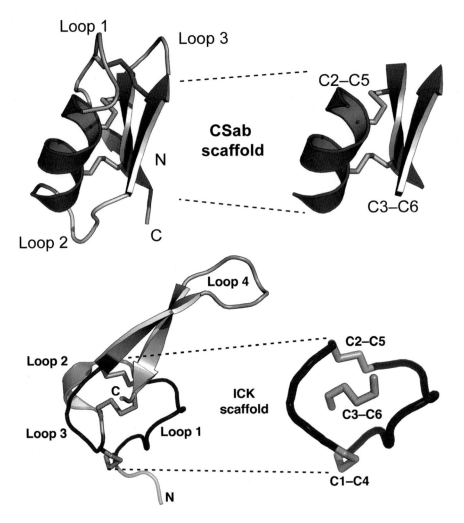

Loop 1

Loop 3

CSab
scaffold

C2–C5

C3–C6

N

C

Loop 2

Loop 4

Loop 2

ICK
scaffold

C2–C5

C3–C6

C

Loop 3

Loop 1

C1–C4

N

A comparison of the 3-D structures of a typical spider and scorpion venom toxin: top, charybdotoxin, from the death stalker scorpion, *Leiurus quinquestriatus*, and below, an ICK-type peptide toxin from the funnel-web spider, *Hadronyche versuta*. The core of the charybdotoxin (enlarged on right) consists of a helix and sheet (blue) stabilized by two disulfide bonds (orange). In the ICK, the cystine knot (enlarged on right) consists of a ring (blue) stabilized by two disulfide bonds (orange), and pierced by the third disulfide bond to create a pseudo-knot, one of the most widespread protein folds in Nature. These two compact and stable toxins have each evolved from a different type of pathogen-defence molecule (defensins).

where 'cystine' (as opposed to cysteine) is not a typo but indicates the formation of disulfide bonds. The knot refers to the characteristic pseudo-knot that consists of a closed ring formed by two of the disulfide bonds and their intervening peptide backbones, and a third disulfide bond that pierces this ring to stabilize the tail of the peptide. This knot is one of the most widespread protein folds in Nature, and has been recruited into the venoms of at least nine animal lineages, including assassin bugs, remipede crustaceans, cone snails and sea anemones.

Although extensive crosslinking by disulfide bonds provides stability to the folded peptide, it does not contribute to solubility *per se*. In other words, a cysteine-rich peptide may be stable, but is still poorly suited to functioning as a toxin if it is not soluble. Interestingly, centipedes in the genus *Scolopendra* and spiders in the genus *Tegenaria* have both independently weaponized a peptide hormone into a venom toxin. In both cases this happened by the peptide losing a protein domain that not only removed its hormonal activity, but the loss of which also makes the new toxin peptide much more soluble.

Expanding a chemical arsenal

Evolution of new functions (what the toxin does on the molecular level) and roles (the biological role of the toxin for the venomous animal) is not just central to recruitment of proteins and peptides into venoms, it is one of the principal drivers of toxin evolution. Proteins and peptides evolve new functions by accumulating changes in the amino acid chain. These changes are often largely the result of the interaction between two opposing selective forces: purifying and diversifying selection. These selective forces affect whether or not random genetic changes (mutations) are lost or kept in a population.

Purifying selection results in the conservation of the amino acid sequences of proteins over evolutionary time. Most amino acid changes to a protein are detrimental to its function, and such mutations are therefore quickly weeded out by purifying selection. This is why many of the housekeeping proteins mentioned previously are so evolutionarily conserved between species. Because most changes would impair the function of these proteins, most species have retained the ancestral form of these proteins. Any mutations disrupting their function can have huge negative effects on the organisms. One way of alleviating this barrier to evolutionary change is to duplicate the gene that codes for the protein in question so that one copy is

result in temporary disarmament, with the fearsome spider becoming a plump, juicy, defenceless meatball. The speed of refuelling venom glands varies considerably between taxa. It can take theraphosid spiders – a group of mostly large and hairy spiders popularly known as tarantulas or bird-eating spiders – as long as 85 days to finish reloading their weapons. In contrast, the tiny caterpillar-killing parasitoid wasp *Bracon brevicornis* can reload its glands in less than three hours. The North American centipede, *Scolopendra polymorpha,* regenerates about 85% of its venom volume within two days.

The great cost of venom, direct or indirect, is also indicated by the phenomenon of 'venom metering'. Snakes can control the amount of venom they release so they do not use more than is necessary. A larger, more dangerous, or more agile prey receives a larger dose of venom to ensure it is subdued effectively. Although venom metering has been poorly studied in spiders, researchers know that they are also able to make decisions about when and how much venom to deliver. The Central American hunting spider, *Cupiennius salei,* injects more venom into

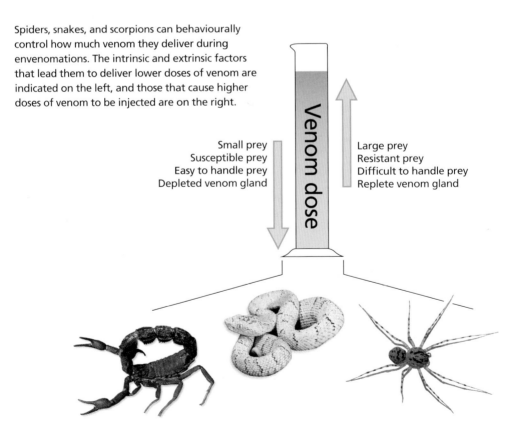

Spiders, snakes, and scorpions can behaviourally control how much venom they deliver during envenomations. The intrinsic and extrinsic factors that lead them to deliver lower doses of venom are indicated on the left, and those that cause higher doses of venom to be injected are on the right.

Small prey
Susceptible prey
Easy to handle prey
Depleted venom gland

Venom dose

Large prey
Resistant prey
Difficult to handle prey
Replete venom gland

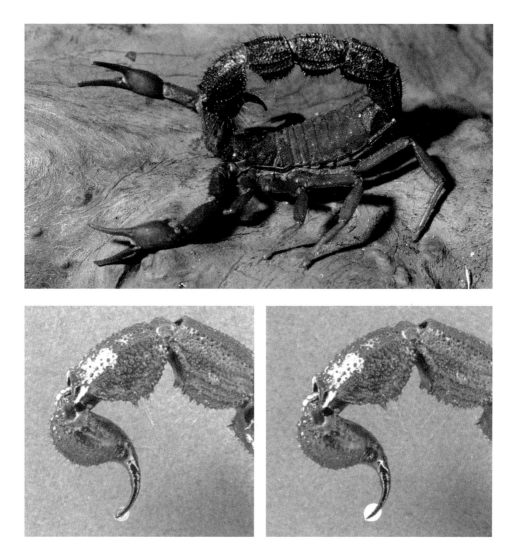

TOP The African thick-tail scorpion, *Parabuthus transvaalicus*, meters its defensive stings depending on the threat level. In a high threat situation it is less likely to deliver a dry sting, and each sting also injects a larger volume of venom than in under low threat conditions.

ABOVE The first sting of *P. transvaalicus* is often a transparent and metabolically inexpensive prevenom (left). After the prevenom is depleted, or if the threat level is high enough to require a large volume of venom, it secretes a more protein-rich milky venom (right).

larger prey. When dealing with prey that has body armour, such as beetles, and the armour prevents the spider from injecting venom into an optimal location, the spider also increases the dose. Remarkably, *C. salei* can also distinguish prey that is more or less sensitive to its venom (crickets and cockroaches, respectively) purely on the basis of smell, and choose the more vulnerable prey when its venom glands are depleted.

Venom is not just metered during prey capture. Snakes, scorpions and spiders also show defensive venom metering. Females of the western black widow spider, *Latrodectus hesperus*, for example, can adjust the amount of venom in their defensive bites depending on the threat level. Squeezing the spider's body provokes her to deliver almost twice the dose of venom than if her leg is pinched. Even so, more than half of her defensive bites contain no detectable venom at all. The defensive bites of some snakes and the defensive stings of scorpions can likewise be cheap shots. The South African thick-tail scorpion initially stings predators with a transparent prevenom that is poor in proteins, but rich in pain-inducing potassium ions. The scorpion releases its milky venom, which is more protein-rich and toxic – but also more expensive to produce – only when its pre-venom fails to do the job.

The metabolic cost of venom has important consequences for the evolution of its toxins. As discussed previously in this chapter, the functional diversification of toxins relies largely on multi-gene families evolving under diversifying selection. However, this is an expensive process because of the increased number of gene copies that likely encode non-functional proteins. Venom evolution is therefore thought to depend on the interaction between two opposing forces: the drive towards increased potency through evolution by diversifying selection, and the conservation of potency under the influence of purifying selection. One recently proposed hypothesis suggests that the trade-off between the cost and benefit of growing toxin arsenals results in a two-step mode of evolution. Toxins first rapidly diversify into new functions by diversifying selection, but are then conserved by purifying selection once the venom has become sufficiently potent. This phenomenon is apparent when comparing differently aged venomous lineages. Younger lineages, such as cone snails, which originated and diversified a mere 50 or so million years ago, tend to have more toxins showing signs of strong diversifying selection. In contrast, toxins from ancient lineages, such as centipedes and cnidarians, which diversified over 400 and 600 million years ago, respectively, are encoded by genes that generally show strong signs of purifying selection.

The banded sea krait, *Laticauda colubrina*, feeds almost
exclusively on eels. This predation pressure has resulted in
eels in the genus *Gymnothorax* evolving a high degree of
resistance to the sea krait's venom where the species co-occur,
but *Gymnothorax* species that do not co-occur with the snake
remain sensitive to its venom.

A uloborid spider, *Hyptiotes paradoxus*, doing what it does
best: tightly wrapping prey in a silken shroud.

The cost of venom does not only affect how venoms evolve, it can also cause a
loss of toxins. Some specialist venomous predators that have evolved to feed only
on a limited number of species, have 'streamlined' their venoms by losing toxin
types that are no longer useful. This removes the cost of producing obsolete toxins.
Sea snakes, for example, have relatively simple venoms compared to their terrestrial
relatives, with the venom consisting of only a few neurotoxins that effectively
immobilize their fish prey. In a fascinating case of convergent evolution, the venoms
of sea kraits, *Laticauda* spp., which are related to sea snakes but independently
adopted a marine fish-eating lifestyle, have undergone the same streamlining.

In fact, the venom of sea kraits is so similar to that of sea snakes that it can be effectively neutralized by sea snake antivenom.

Venom is such an expensive commodity that it is ditched entirely if its ecological importance disappears. For example, when the marbled sea snake, *Aipysurus eydouxii*, switched to a diet consisting exclusively of fish eggs it didn't simply lose a few toxin types. It lost its venom, fangs and even its venom gland. Similarly, spiders in the Uloboridae family catch prey using only their silk, and have subsequently lost their venom glands. Instead, they regurgitate digestive enzymes to dissolve their immobilized prey, which they wrap tightly in a silken burrito of death, before ingesting it through their straw-like mouths.

You are what you eat

Not surprisingly, the ecological roles played by a venom dictate how its toxins evolve. In predatory roles the high cost of venom means that there is a strong drive for optimizing its potency. And in the quest for potency, it is the target that selects the toxin. In other words, the way toxins evolve is governed by their physiological targets in the envenomated animal. For example, toxins from venoms across the animal kingdom have evolved to target almost all steps in vital physiological pathways from nerve signals to blood coagulation. This stunning convergence extends to the specific receptors involved in each of these pathways. At the time of writing, for example, there are over 400 peptide toxins that are known to target sodium ion channels (see chapters 3 and 4), including 182 scorpion toxins, 116 spider toxins and 55 sea anemone toxins. Because sodium ion channels play a key role in initiating nerve impulses, it is no great surprise that they have become a bull's eye for paralyzing venoms.

However, the optimization of potency often comes at the price of selectivity. Although proteins of physiological pathways targeted by toxins tend to be similar across animal lineages, they nevertheless nearly always contain some differences. As we saw in chapter 4, for example, the mutation of just a single amino acid in the sodium ion channels of the German and American cockroaches makes all the difference when it comes to the ability to resist some spider neurotoxins. Venomous animals, therefore, have to tailor their venom to the intended victim. The toxicity of venoms from species of saw scaled vipers, *Echis* spp., against scorpions, for instance, varies according to the proportion of scorpions in the diet of each species. Saw scaled viper species that feed more on scorpions have venoms that are more potent

The venom of the Eastern brown snake, *Pseudonaja textilis,* contains toxins that are powerful disruptors of blood clotting. One of these, called textilinin-1, inhibits enzymes that dissolve blood clots. It is currently being developed into a drug for the treatment of bleeding.

against scorpions than the venoms of species that focus on other prey. Similarly, the venom composition in populations of the Malayan pit viper, *Calloselasma rhodostoma*, depends on the prey types that make up its diet. *C. rhodostoma* is a generalist predator that feeds on amphibians, reptiles, birds and mammals, and it is the relative proportion of each prey type in its diet – not the geographical distribution or evolutionary relationship of species – that best explains the patterns of venom variation that can be observed across the range of this species.

Further supporting the idea that venomous animals tailor their venoms towards their targets, several venomous animals change their venom composition according to shifts in their diet that occur as they grow and mature. One such animal is the creatively named and decidedly brown-coloured Eastern brown snake of eastern Australia, *Pseudonaja textilis*. This snake begins its life as a reptile specialist with a venom typical of other Australian reptile specialists, dominated by peptide toxins with mainly neurotoxic effects. Upon passing a length of about 60 cm (2 ft), however, the diet of *P. textilis* shifts to include a large proportion of rodents. This leads to an increase in the venom of the extremely potent blood-clotting complex, discussed in chapter 4, that is made up of a weaponized clotting factor and its cofactor. This complex is unique to the venoms of Australia's most infamous cousins, the brown snakes, *Pseudonaja* spp., and taipans, *Oxyuranus* spp., where it was a key evolutionary adaptation that allowed a dietary switch to mainly rodent prey. It was also, in terms of venom LD_{50} values (see chapter 3) and average venom yield, a switch that made these the deadliest snakes on the planet in terms of venom toxicity.

One issue with optimizing venom for predation on one organism is that using it for defence against another organism may become less effective. Alternatively, primarily defensive toxins may be included when envenomating prey and vice versa. In both cases the result is that the cost of venom use increases, regardless of how carefully metered it is. To get around this issue, cone snails have opted for another way of saving on the energy expenditure of venom. Instead of just metering the amount of total venom that is secreted, they have differentiated their venom glands into two sections that produce distinct toxic concoctions used exclusively for defensive or predatory purposes (see chapter 3). Several toxins have also diversified within these functional groups, forming defence- and predatory-specific toxin lineages. Interestingly, this functional diversification of defence-specific toxins has been proposed to underpin the dietary switch of cone snails. Toxins first used in the defence against predatory molluscs (octopuses) and fish, enabled a switch from the ancestral diet of worms to feeding on molluscs (snails) or fish, respectively. Cone snails have thereby found a way of turning the evolutionary constraint that is the metabolic cost of venom into an evolutionary driver.

Evolution of toxin resistance

Streamlined, potent venoms have evolved to minimize the metabolic costs of producing them. But animals whose lives frequently intersect with venom can evolve resistance to these finely honed weapons. Before we go on to explore the evolution of resistance, however, we need to address the elephant in the room: what about the venomous animals themselves? Surely they must be resistant to their own venoms, given that their lives are completely intertwined with venom? In cases where venom is used solely for use on members of the same species, a certain level of resistance almost certainly evolves. Such an arms race between generations is what is thought to underlie the extreme pain-producing abilities of the platypus (see p.63). In most cases, however, self-immunity remains a poorly studied area of venom research. Among venomous snakes, for example, the Australian death adder, *Acanthophis antarcticus*, has been reported to be immune to, but also die from, envenomations by members of the same species. This is also the case for other elapid snakes, as well as rattlesnakes, despite the latter having proteins in their blood that are able to potently inhibit the activity of their toxins. In addition, several spiders are well known to cannibalize their own species, a task that necessitates the action of venom.

There are, however, certainly examples of animals that are not just resistant but apparently even immune to their own venom. Applying venom from the red spotted assassin bug, *Platymeris rhadamanthus*, to dissected hearts from the same species, for example, has no effect. In contrast, the same venom applied to the same preparations from a cockroach immediately and violently stops the heart by causing a massive contraction. The aptly named scorpion *Androctonus australis* (which translates to 'Southern killer of men') also appears to be immune to the potent neurotoxic effects of its own venom. While its venom has a strong neurotoxic effect on the nerves of crayfish, nerve preparations from *A. australis* remain unaffected at the same concentrations. This is perhaps unsurprising since males deliver sexual stings to females during mating, without killing her.

While self-immunity in venomous creatures still poses many unanswered questions, the evolution of resistance is well documented in species that interact with the venoms of other species. Animals acquire resistance in three main ways: toxin inhibition, modifying the target of the toxin and repurposing one of the toxins in the venom to counter the actions of the others.

Resistance is best characterized in animals whose lives frequently cross with venomous snakes. In rattlesnake venom, the aforementioned SVMPs are among the most important components. Being proteases, an effective way of rendering these toxins harmless is by inhibiting their enzymatic activity. Resistance against many viper venoms is therefore achieved through SVMP inhibition by proteins found in blood serum. These proteins are all produced by the innate immune system, and they function by binding irreversibly to the toxin and rendering it unable to perform. Ground squirrels on the West Coast of the USA that coexist with large populations of the Northern Pacific rattlesnake, *Crotalus oreganus oreganus*, for example, have evolved high resistance to the effects of SVMP largely through the action of serum SVMP-inhibitors.

Like *Crotalus oreganus oreganus*, the Great Basin rattlesnake *Crotalus oreganus lutosus*, is a subspecies of the Northern Pacific rattlesnake that also primarily feeds on rodents.

Another way of acquiring resistance is by modifying the target of toxins. Although this is probably the most commonly investigated mechanism of gaining resistance to venoms, it has only been documented a handful of times. Interestingly, none of the cases of acquiring venom resistance due to changing the toxin target are due to the interaction with a venomous predator, but instead involve venomous prey. The Egyptian mongoose, *Herpestes ichneumon*, and the honey badger, *Mellivora capensis*, which both eat venomous snakes, have independently modified

Although it is unknown if it is resistant to snake venom, the slender mongoose, *Galerella sanguinea*, is known to prey on venomous snakes, such as this boomslang, *Dispholidus typus*.

the primary target of the snakes' neurotoxins, namely a neurotransmitter receptor known as the muscular nicotinic acetylcholinesterase receptor, which mediates signals between nerves and muscles. The mongoose has opted for the same solution as the cobras on which it preys. By adding a sugar chain to one of the amino acids near the toxin-binding site, the neurotoxin is physically blocked from binding to the receptor. If you imagine the target on the receptor to be like an egg cup, and the toxin to be like an egg that snugly fits the cup, the sugar chain is like a sugar cube in the cup that prevents the egg from fitting. The honey badger has modified its neurotransmitter receptor in a different way, by substituting several amino acids in the toxin-binding site with amino acids carrying an opposite electric charge. This repels the toxin from its target in a manner similar to the way in which the corresponding poles of two magnets repel each other.

The third way of gaining resistance to venoms is a particularly intriguing one, namely repurposing toxins to block the effect of the venom. Contrary to their cuddly sounding common name, grasshopper mice, *Onychomys* spp., are ferocious

A grasshopper mouse feasting on the spoils of
its battle with a bark scorpion.

predators, and use this strategy of gaining resistance to feed on highly venomous
bark scorpions, *Centruroides* spp. Although bark scorpions sport an incredibly painful
sting, which they achieve by activating a type of sodium channel called $Na_V1.7$ that
initiates pain impulses, grasshopper mice that feed on them are largely unaffected.
Instead of having evolved mechanisms to prevent toxins from activating $Na_V1.7$,
grasshopper mice have modified another type of sodium channel, called $Na_V1.8$, so
that it is targeted by the same venom. $Na_V1.8$ is needed to sustain the nerve impulses
produced by $Na_V1.7$, but because the scorpion venom blocks $Na_V1.8$, any impulses
generated by $Na_V1.7$ are not transmitted any further. Thus although turning more
receptors into targets of toxins may seem like an odd strategy for generating venom
resistance, the venom essentially switches off any pain signals produced by one
target by triggering another target to block them. This means no pain signals make
it to the brain. In a fascinating case of adaptive evolution, grasshopper mice have
turned the pain-inducing venom of the bark scorpion into a painkiller.

Chapter 6

Cultures, Cures, Quackery and Cosmetics

In December 2014, two employees of a pharmaceutical company visited the Natural History Museum in London to meet with museum staff. One of the authors had arranged several specimens on a table to give the visitors some idea of the amazing diversity of venomous life on Earth. But as soon as the guests entered the room, one of them rebounded with disgust, turned on her heels and retreated to the hallway. A specimen had struck sudden fear in her heart. The author's protestations that there was no danger at all, that all the specimens were assuredly dead, either taxidermied or hermetically sealed in jars of alcohol, fell on deaf ears. Our visitor categorically refused to enter the

OPPOSITE A 3D scan of the skull of the Gaboon viper head, showing the large venom fangs, from the collections of the Natural History Museum, London (right), that struck fear into the heart of an official visitor.

This is all too human behaviour. Much that frightens and disgusts us also exerts an irresistible pull. We slow down and gawk at traffic accidents. We safely peek at horror movies from behind our pillows. In bygone ages, gladiatorial fights were the acme of entertainment for rulers and laity alike. We watch reality TV shows about first aid responders and emergency rooms. Jailed rapists receive love letters and marriage proposals. We put books about serial killers on the bestseller lists, and we are addicted to *Crime Scene Investigation*. We made Stephen King rich.

Fear also begets respect. We feel awed by power, and the power of venom has accompanied us during the whole of our evolutionary development. The ancient Egyptians worshipped deities that were symbolized by sacred venomous animals, including scorpions, snakes, centipedes and water scorpions (aquatic relatives of assassin bugs). Snakes were associated with a number of gods, both good and evil. In addition to protecting the dead, the Egyptian centipede god Sepa was thought to protect people from deadly and venomous creatures, especially snakes. The earliest known depictions of venomous arthropods as symbols of protective deities grace the mortuary tomb of pharaoh Sahure from more than 4,000 years ago. The pillars of the oldest known temple site in the world, the 11,000-year-old Göbleki Tepe in Turkey, are likewise adorned with etchings of spiders, snakes and scorpions. And the worship of snakes pervades the long history of civilization in many parts of the world.

The more or less formalized representation of venomous creatures as religious symbols is only a tiny facet of the variety of forms that they assume in general culture and mythology. They adorn the labels of beer brands and national flags, they are protagonists in movies and books, they feature in art throughout human history recorded anywhere from canvas to the human skin, they inhabit the bedrooms of exotic pet enthusiasts, and they haunt us in our dreams. But the cultural ubiquity of venomous animals is not restricted to the realm of the symbolic. The damaging power of venom has been unleashed deliberately. Clay pots filled with snakes or scorpions were hurled at the enemy. Arrows adorned with mixtures of poison and venom have been used for the hunting of both man and beast. Some Amazonian tribes, for instance, create a diabolical type of ammunition by dipping their hunting arrows in a mixture of the painless, paralytic plant poison curare and the incredibly painful venom of bullet ants. Nevertheless, the use of venom to harm is rare compared to the other ends to which human ingenuity has fashioned it.

In this chapter we will explore both sides of the world of venom, the dark and the light. First, we will delve deeper into the fear surrounding venomous animals,

The Mexican flag features an eagle grabbing a rattlesnake
with beak and talons.

with a focus on venomous snakes and the treatment of their bites. Second, we will
see how the deliberate use of venom and venomous animals ranges from traditional
medicine to modern cosmetics and rites of passage in different cultures. And finally,
we will look at the proven value and future promise of harnessing the power of
venom for the development of new pharmaceuticals.

A universal fear

Why can snakes trigger fear and panic in so many people in largely snake-free
urbanized societies? Why are Dutch people so afraid of snakes when the only
potentially dangerous snake that they are likely to encounter in the wild – and they
have to search hard if they are to be so lucky – is the European adder, *Vipera berus*? To
answer these questions we have to broaden our perspective beyond the Netherlands.

In 1991, anthropologist Donald Brown published a remarkable book titled
Human Universals, which reported the results of a comprehensive ethnographic
survey of features of human culture, society, language, behaviour and mind that are
shared by all peoples that have ever been studied. The result was a list of hundreds

of traits that represent universal facets of human nature. But buried amidst music, incest taboos, myths, hospitality and customary greetings is only one universal trait that explicitly refers to a living thing that is not human: fear of snakes. And not only is our ophidian fear universal, it is innate as well.

An innate fear of snakes has been documented in Old World and New World monkeys as well as apes. Humans, too, come into the world with an innate predisposition to becoming afraid of snakes. From the age of about five, children grow wary around snakes, and only one or two unpleasant experiences – a sudden wriggling motion in the undergrowth, a lunging snake in a movie – can trigger a deeply rooted, lifelong fear that is very difficult to overcome. In some unfortunate people this creates the emotional equivalent of anaphylactic shock. Sensitized by one or two bad experiences they develop ophidiophobia, and fall prey to a disproportionate and debilitating fear when faced with a snake. Just as some venoms can cause anaphylaxis by triggering an overreaction of the immune system, certain venomous animals, especially snakes and spiders, can trigger phobias by an overreaction of the emotions. But why would we be genetically predisposed to become averse to snakes, and why does it take so little to make this fear pathological?

Our fear of snakes betrays our bestial origins. Our ancestors must have lived in environments where the threat of dangerous snakes was so significant that it was advantageous to be able to detect and avoid them before they could do any harm. Individuals who were better able to avoid dangerous snakes had a better chance of surviving and raising a family. If this ability had any genetic components, and if it led to a great enough reproductive advantage, aversion to snakes would spread through the population until it became fixed as a universal trait. Today there are only a handful of species of non-venomous snakes that are large enough to be dangerous to large-bodied primates such as us. The vast majority of snakes living today that can cause serious injury or death are venomous. Over the past several million years, our ancestors faced precisely the same situation. Hence, venom is the likely selection pressure that stood at the cradle of our universal fear of snakes. The selection pressure of venomous snakes may have produced more than innate fear. Anthropologist Lynne Isbell has proposed the fascinating theory that the improved ability to detect and avoid venomous snakes has been a crucial driver of the evolution of increasingly complex and sophisticated visual systems and brain circuitry in primates. Among other things, this process has resulted in the evolution of neurons in the brains of monkeys that selectively respond to images of snakes.

The impact of snakebite

Dangerous venomous snakes occur on all continents except Antarctica, and they are common in many of the environments in which traditional societies continue to survive. In his book *The World Until Yesterday*, Jared Diamond summarized the main causes of accidental death and injury for seven traditional societies from Africa, South America, Asia and Oceania. In four of these societies, bites by venomous snakes rank among other top causes such as infected insect bites and falling out of trees. But for one group of South American rainforest dwellers, snakebite trumps all other dangers.

In 1996, anthropologists Kim Hill and Magdalena Hurtado published an in-depth study of the lives of Ache Indians, a small population of indigenous people living in the rainforests of eastern Paraguay. The Ache are hunter-gatherers and their homelands swarm with venomous creatures. Before they started living in reservations they were nomadic, but they are still forest hunters. Stings and bites by spiders, ants, bees and wasps are an almost daily occurrence. As Hill and Hurtado write in their book *Ache Life History*, 'Forest camps are constantly interrupted by the cry of some child who is learning the hard way about which insects to avoid.' But although arthropod envenomations were an ever-present nuisance for the Ache, the incidence of snakebite was truly stunning.

During their time with the Ache Indians, Hill and Hurtado frequently saw Ache kill snakes, as they did many times themselves. They witnessed about half-a-dozen snakebites, some of which resulted in paralysis or destruction of the bitten limb. For Ache men who hunt in the forest for seven hours every day, snakebites were unavoidable and Hill and Hurtado found that most adult men had been bitten at least once. Indeed, snakebite was the most important cause of accidental death and injury in Ache society. Astonishingly, snakebites kill 14% of adult Ache men. One day, Hill almost experienced the most common scenario of snakebite. While scanning the canopy for game he almost stepped on a snake and narrowly avoided being bitten only because someone yelled a warning.

Snakebites are responsible for 6% of all Ache deaths. This represents an extreme position along a broad global spectrum, but inconsistent reporting makes it notoriously difficult to get reliable estimates of the incidence of snakebite. Many published figures are likely to be serious underestimates. India is generally considered to be the snakebite capital of the world in terms of absolute numbers of snakebites per year. A massive study in India, known as the 'Million Deaths' study, found that 0.47% of all deaths between 2001 and 2003 were caused by snakebite, corresponding to almost 46,000 deaths. But although India may top the snakebite mortality ranking in absolute numbers, other parts of the world have higher, and sometimes much higher, mortality rates standardized by the size of the population. For India, this metric is 4.1 snakebite deaths per 100,000 people per year. In the rural town of Kilifi in Kenya it is 7. In the Burdwan district of East India it is 16, while in the West African savannah it is between 4 and 40. In the Terai region of Nepal it is 162, which makes it the highest snakebite mortality rate in Asia. This enormous mortality rate puts it well within the range of that for cardiovascular disease in the UK, one of the country's leading causes of death. It is, however, strongly dependent on socio-economic factors. At 0.01, the rate of snakebite mortality in the USA is more than 400 times lower than for India as a whole, and more than 16,000 times lower than for Nepal's Terai region. In the US you are over 10 times more likely to die of a bee sting than of any other type of envenomation, be it by snake, spider or scorpion.

A man handling a king cobra, *Ophiophagus hannah*, in the Western Ghats of southern India.

The chances of being bitten by a snake can differ dramatically across small geographic distances and habitat types, and certain occupations pose much greater risks than others. Snakebite is an occupational hazard for many farmers, plantation workers and herders in the tropics, and in some places where venomous sea snakes are common, for fishermen as well. The high incidence of snakebite in Nepal's Terai region, for instance, is a result of a tropical climate, lush vegetation and a high density of rodents alongside a high human population density, with most bites happening in the rainy season when agricultural work is at its most intense.

Snakebite is responsible for many deaths and the untold misery of many survivors who end up with chronic physical handicaps. Recent conservative estimates suggest that globally there are around 5 million snakebites every year. About half of these are thought to result in envenomations, the treatment of which requires 400,000 amputations. About 100,000 of the victims die, with most of these living in Southeast Asia and sub-Saharan Africa. The physical disfigurement of survivors often has severe psychological, social and economic consequences. Disfigured girls may not be able to get married and start a family. Injured limbs or eyes – caused by spitting cobras

Snake species that cause large numbers of bites and fatalities are: above left, *Echis* spp. (saw-scaled vipers) in Africa, above, *Naja* spp. (cobras) and *Bungarus* spp. (kraits) in Asia, and left, *Bothrops asper* and *B. atrox* (lance-headed pit vipers) in Central and South America.

– can make it very difficult or impossible for the victim to make a living. The World Health Organization has therefore recognized snakebite as a neglected tropical public health issue since 2009, as it harms more people than several other neglected tropical diseases, such as leishmaniasis (a disfiguring disease caused by a parasitic protozoan), leprosy and yaws (chronic, disfiguring bacterial infections).

Unfortunately, the resources made available by governments and donors to tackle this problem are in many places wholly incommensurate with the medical, social and economic damage caused by the bites of snakes. Many lives are ruined or lost because of the lack of emergency care and the scarcity of effective and affordable antivenom. In some places the costs of a round of antivenom treatment can bankrupt victims. To receive antivenom treatment for a snakebite may cost someone in sub-Saharan Africa somewhere between $500 and $600. For the many people who earn less than $1 a day, this is an insurmountable burden.

The people who are most at risk of snakebite are typically also those who are most frequently envenomated by other creatures, notably scorpions. Being a farmer anywhere in the tropical and sub-tropical regions of the world is not for the faint hearted. Scorpions sting an estimated 1.2 million people a year, causing more than 3,200 deaths. People most at risk are those in Mexico, where the incidence of scorpion stings in the state of Morelo, located just south of Mexico City, is a stunning 2,050 stings per 100,000 people per year. But being a soldier is the job in which you can be at greatest risk of running into the business end of a scorpion. In 1991, the corresponding statistic was 2,400 for American troops deployed in Operation Desert Shield during the Gulf War.

The need to neutralize: fighting fire with smoke

Although insects, spiders, scorpions and jellyfish together deliver far more venomous stings and bites to humans than snakes do, they cause far fewer deaths. The ever-present danger of snakebite that has accompanied the evolution of the human race has become distilled in our uniquely universal fear of snakes. It should therefore come as no surprise that recipes for the treatment of snakebite go back to the earliest known medical texts. Snakebite remedies also exist in great abundance in traditional cultures that lack written records. But the fate of a snakebite victim, for good or ill, depended on when in history or where on Earth he or she received treatment. In some cases a so-called 'cure' could make a bad thing much worse.

Before their lives became centred around reservations, the Ache took little action when they were bitten by a venomous snake. They would simply grit their teeth and hope they would survive without debilitating injuries. Living in groups without shamans or healers, they might have attempted to blow some smoke from a fire over the bite wound, but that's where treatment would end. Many other cultures, however, have taken more active approaches to neutralizing the destructive power of venom. The Brooklyn Museum of Art holds an Egyptian papyrus that is around 2,500 years old. This medical papyrus details a diversity of spells and remedies that were recommended for the treatment of snakebite, including lancing of the wound. The famed encyclopaedia *Historia Naturalis*, compiled by the Roman writer Pliny the Elder 2,000 years ago, advised that snakebite should be treated by anointing the wound with a concoction of fresh sheep turds cooked in wine, although he felt that the application of a bisected rat to the wound could be equally effective.

Treating snakebite with potions derived from plants, animals and minerals is ubiquitous in medical traditions throughout history and across the world. To get

The title page of Pliny the Elder's *Historia Naturalis*. The first edition was published in Venice in 1469 and included advice about the treatment of snakebite with fresh sheep turds.

A CURIOUS CONCOCTION

Medical texts written in medieval times contain a litany of concoctions and decoctions intended for internal and external use to treat snakebite. One of the most enduring and widely used concoctions was known as Theriac. Considered an all-purpose remedy, it was prescribed for an astonishingly long period – from the first to the nineteenth century – for afflictions as diverse as smallpox and snakebite. It could be swallowed as well as applied externally, which might have been favoured by those in need of treatment given that among its 70 or more ingredients from all corners of the world were opium, viper flesh, asphalt, gum Arabic, cinnamon, balsam wood, castoreum (a secretion that beavers use for territorial scent marking) and ground up mummy.

An 18th century theriac jar from Italy.

some idea of what a traditional system of healing looks like in a global hotspot for snakebite, let us have a look at Ayurvedic medicine. Ayurveda is an ancient medical tradition with roots in India that go back 3,000 years. An especially important strand of traditional healing that fed into Ayurvedic medicine, and that focused specifically on snakebite, is captured in the Garuḍa Tantras, an early branch of Hindu scripture. The tantras are a religious medical manual, and are named for the bird deity Garuḍa, arch enemy of all things ophidian and venomous. The recent publication of an English translation of some medieval Garuḍa Tantras by the historian Michael Slouber shows what divinely ordained snakebite treatment looks like. This work, titled *Kriyākālaguṇottara*, ostensibly represents a conversation between the Hindu god Shiva and his son Skanda, in which the father relates his medical teachings. Depending on the symptoms, Shiva advises the use of decoctions that are predominantly different mixtures of various types of plants and plant parts that the victim needs to consume at the same time as they are being applied externally.

The type of mixture that the *Kriyākālaguṇottara* advises depends upon how deep the venom is thought to have penetrated into the patient. There is a sudden shift from plant matter to the use of animal and chemical ingredients as the venom gets deeper. When all else fails Shiva advises the patient be treated with a mixture of pigeon eyes with yellow and red arsenic, again for simultaneous internal and external use.

Although treatments such as this were recommended over a thousand years ago, very similar ones remain important in Indian traditional healing today. A classic Ayurvedic text that offers extensive advice

The bird diety Garuḍa carrying a jar of amrita (elixir of immortality) that he intended to deliver to the serpents who kept his mother imprisoned, in exchange for her freedom.

on the treatment of snakebite that has remained popular today since it appeared in the 1930s is titled *Prayoga samuccayam*. It suggests that unconscious patients can be brought to by having them smell a mixture of the first dung of a calf ground into the urine of a goat. Although this may indeed be efficacious, one wonders about the ease of obtaining such ingredients in an emergency. And if someone who has been bitten by a viper starts bleeding from his or her hair follicles, the *Prayoga* advises the healer to mix fried powder of the moringa tree with ghee, and to rub it all over the patient.

Traditional snakebite treatments such as these contrast starkly with the approaches of modern medicine. But the greatest rift between Ayurveda and many other forms of traditional healing on the one hand, and Western medicine on the other, is the role of spiritual and mystic elements in the diagnosis and treatment of envenomations. Traditional treatment can involve elaborate rituals that can take years to master, as well as the chanting of mantras that are claimed to be able to, for instance, cause the constriction of blood vessels, thereby slowing the spread of venom through the body.

Ayurvedic approaches to snakebite diagnosis are equally remarkable from the perspective of our modern understanding of how venom attacks bodies. If the victim doesn't know what snake has bitten him or her, a mixture of seeds and leaves, oil and chemicals is mixed and purified and then eaten by the victim. The pungency or sweetness of the taste then reveals what snake has been the culprit and how bad the envenomation is. Not only this but the identity and behaviour of the messenger who informs the healer that someone has been bitten, are thought to be significant for the medical prognosis.

The *Prayoga samuccayam* states that an experienced Ayurvedic physician should even be able to infer the type of snake that the victim was bitten by on the basis of the location of the messenger in his office. Moreover, the locations of the messenger and healer in the room can similarly reveal tell-tale signs about where and how badly the patient has been bitten. In fact, the author of the *Prayoga*, Kochunni Thampuran, is said to have been able to prepare the appropriate medications for snakebite solely on the basis of astrological calculations before victims even reached him.

Other treatment options that continue to have a place in Ayurvedic and other forms of traditional healing include using one's mouth to try and suck the venom out, pressing a snake stone – often a charred piece of animal bone or a stone created by baking pulverized pebbles together with leaves from medicinal plants – against the bite wound to absorb the venom, applying a tourniquet to restrict the spread of venom, cutting the bite wound to let the venom bleed out, inducing the victim to vomit and cauterization (the burning of a wound to sterilize it and staunch any bleeding).

A depiction of a man being attacked by a snake, from Jacob Meydenbach's 1491 *Hortus sanitatis*, the first natural history encyclopedia that describes the medicinal value of plants, animals and minerals.

This mix of mysticism and materialism in the treatment of snake envenomation is still very widely practised all over the world, particularly where Western medicine isn't readily available. But even in countries where hospitals with emergency care facilities and antivenom are available, snakebite victims often seek out traditional healers first, or even exclusively. For example, a 2009 report on snakebite in Cambodia, commissioned by the World Health Organization, reveals that most snakebite victims do not attend hospitals, but instead consult traditional healers. Many who do eventually go to hospital will have sought traditional help first, and some patients (seven out of 22 snakebite victims over a period of two years in Kampong Thom Provincial Hospital) discharge themselves from emergency care to return to traditional healers.

The results of failing to receive proper emergency care can be life threatening. Tourniquets can cut off blood flow and result in gangrene or exacerbate tissue destruction by concentrating toxins in the bite area. This is a particular risk with the enzymatic venoms from vipers. When envenomated with haemorrhagic venom, creating incisions in the wound can cause unstoppable bleeding. Trying to absorb the venom with a snakestone, trying to suck it from the wound, or applying non-sterile potions can cause dangerous infections. And even when traditional methods do not directly cause any harm, any delays in getting modern medical care increase the chances of the victim ending up with debilitating injuries or worse.

Yet, the use of traditional methods, including all the examples given here, continue to be discussed and endorsed even in peer-reviewed academic literature that is dedicated to the study of traditional and alternative medicine. This literature often claims that Ayurveda and other medical traditions operate in semi-independence from scientific medicine, and that they deserve respect in their own right because they follow fundamental principles that are distinct from those that underpin scientific medicine. This makes it ironic when proponents of traditional healing appeal to the authority of scientific consensus to vindicate their own methods when the two approaches do converge.

The vast corpus of traditional medicinal knowledge might appear to offer patients a reassuringly broad menu of treatment options incorporating both material and spiritual elements, but in reality this breadth of choice is a millstone around the neck of traditional healing. It betrays a lack of consensus on what treatments work, and the wilful inertia of tradition hampers progress in accepting, delivering and advocating those treatments that have been scientifically proven to be truly effective. Some modern Ayurvedic texts and religious scriptures from millennia ago show such striking similarities in the treatments they suggest for snakebite that it appears they could have been written by the same person, as they indeed may have been. But this is less a sign of success than of stagnation. Traditions deserve respect when they are a force for good, and they do not promote treatments that are actively harmful. But when healers wilfully ignore well-established scientific consensus in the treatment of snakebite, venerable tradition turns into life-threatening quackery.

Tragically, most snakebite victims live in the poorest parts of the world, where traditional medicine is often the only possible balm for their ailments. In these circumstances, advocates of traditional healing of course rightly point out that if it weren't for their services, patients would have nowhere to go. Indeed, traditional healers can provide emotional support to patients, and besides possible beneficial placebo effects their medicinal remedies can possibly help alleviate symptoms such as pain, nausea and inflammation. However, these benefits are counterbalanced and frequently outweighed by the risks associated with the same treatments and delays to receiving modern medical care.

From quackery to cure

In its protracted infancy, modern medicine went through its own experimental phase in the treatment of snakebite. Driven by the all too understandable conviction that to do something must be better than to do nothing, snakebite victims were patched with potions, cauterized, lanced, bled and injected with harmful substances such as ammonia and strychnine. The Brisbane *Courier-Mail* from 8 October 1936 features an article titled 'Boy slapped for eight hours,' which claims that '[e]ight hours of slapping, scolding, and shaking saved the life of Cecil Schultz (11)' after he was bitten by a snake. Survive he did, but it might not be unreasonable to suppose that this was not because of, but despite his treatment. As the scientific understanding of snakebite grew, such treatments were exposed to be either ineffectual or positively harmful. Nevertheless, a 2010 review of 48 websites that offered advice on the pre-hospital treatment of snakebite found that more than half of them still made inappropriate recommendations, including the use of ice and electric shock on the bite wound.

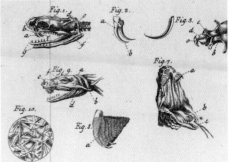

ABOVE RIGHT A banded krait, *Bungarus fasciatus*, skin mounted on paper, part of the Patrick Russell collection of snake skins in the Natural History Museum, London, from the late 19th century.

RIGHT Felix Fontana published his *Traité sur le vénin de la Vipere* in the late 18th century, which included descriptions of many experiments with viper venoms, and snakebite treatments.

Good intentions aren't enough to make for effective treatment. Venom toxins do their damage by crossing the skin barrier, travelling deep inside the body and disrupting the structural integrity and physiological functioning of the victim by fastening onto very specific molecular targets with high affinity. It remains unclear how smoke blown over a bite wound, tantric soundwaves, herbal potions either swallowed or applied externally, or a good bout of slapping are able to intercept and neutralize the destructive power of venom toxins that have been honed by millions of years of evolution to cause rapid and reliable damage.

It is true that hundreds of plants are known for what is sometimes referred to as their anti-ophidian activities, and many of these are used in traditional healing. Extracts from such plants have been shown to have all manner of activities that could be beneficial in the treatment of snakebite symptoms, such as the inhibition of tissue-destroying enzymes. There remains huge potential in the continuing search for natural products such as these, which can have real medicinal value for a host of afflictions, and this remains a cornerstone of scientific and traditional medicine alike. But so far, not a single potion made of plant, animal or mineral ingredients, and not a single chant nor ritual has been shown to be as effective as the timely administration of the only scientifically validated treatment for envenomations: a blood transfusion from a hoofed mammal.

The method of producing antivenom has remained essentially the same for 120 years. A large mammal, often a sheep or horse, is injected with a low dose of venom from a chosen venomous animal. This induces its immune system to form antibodies against the foreign venom components. After receiving several booster shots with venom to increase the amount of antibodies,

Dried herbarium specimens of the goat's foot, *Ipomoea pes-caprae*. In Australian aboriginal medicine it is used as a poultice to treat stingray and stonefish stings.

the animal's blood is collected, and its blood serum, which contains the antibodies against the venom, is isolated. After purification it is stored and ready for use as antivenom. The antibodies in the antivenom neutralize the destructive force of venom by preventing toxins from reaching their targets. To be effective, the antivenom is injected into the veins of an envenomated victim, which ensures that the antibodies quickly and efficiently spread throughout the body, neutralizing any toxins they encounter.

Antivenom, when available, can be a true lifesaver. Unfortunately, it isn't always available where it is most needed. For instance, until 2016 an antivenom known as Fav-Afrique had been used to treat snakebite in Africa, and it was effective against the venoms of 11 species of vipers, mambas and cobras. However, the pharmaceutical company who made it has ceased production, and the last remaining stocks passed their best-before date in the summer of 2016. This is very bad news for farmers in sub-Saharan Africa. But even when Fav-Afrique was available, it was very expensive, so that antivenom treatment could bankrupt snakebite victims at the same time as saving their lives.

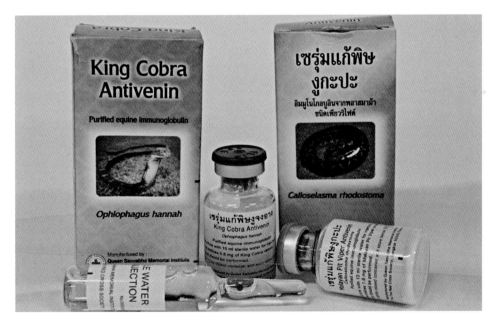

Examples of modern antivenoms that are available for the treatment of snakebite by different kinds of snake.

Although the traditional method of producing antivenoms has saved many lives, it doesn't always do so without problems. The pool of antibodies isolated from a herd of sheep or horses that have been immunized with venom will contain more than just antivenom antibodies. Their blood serum will contain a cocktail of antibodies against all manner of things that the animals may have encountered, such as infections or illnesses. The venom-neutralizing antibodies may therefore only be a fraction of the total amount of antibodies in the antivenom, and indeed, some marketed antivenoms are very weak and need to be administered in high doses to be effective. Another problem is that a substantial proportion of people show adverse reactions to antivenoms, including life-threatening allergic reactions.

To address these problems relating to the cost, availability and efficacy of antivenoms, researchers are trying to design antivenoms in the laboratory that only contain the specific antibodies needed to neutralize a given venom or group of venoms. We are probably on the cusp of a new era of designer antivenom the production of which is made possible by recent biotechnological advances. In order to effectively neutralize a venom you have to know a lot about the toxins it contains, and how they work alone and together to disrupt targets in the victim's body. The paradoxical result is that our scientific understanding of how venom works is advanced enough to be able to develop powerful venom-based drugs for the treatment of a host of afflictions, which are worth billions of dollars. However, the amount of knowledge required about a venom to be able to neutralize it without using the assistance of the immunological genius of a herd of hoofed mammals has taken far longer to achieve.

From traditional healing to beetox and snake oil

Venomous animals and their venoms have been administered to heal the sick, boost the listless and protect the healthy, as well as to prettify the vain, to intoxicate drug users and even to provide entry to the realm of the supernatural. Their perceived power has earned venomous creatures a place in the pharmacopeia of traditional medicine the world over. As we saw earlier, viper flesh was an ingredient of the multi-purpose potion known as Theriac, which was used, among other things, to cure snakebite. The larders of traditional Chinese medicine are overflowing with venomous animals, including snakes, scorpions, antlions, spiders and centipedes. One recipe from a 1998 book titled *The Healing Cuisine of China* is called 'Centipede and Licorice [sic]', and is specifically designed to cure what it calls 'male sexual deficiencies'. It recommends grinding dried centipedes into a powder together with liquorice and anise seeds, and to take half a teaspoon of the mixture twice daily. Southeast Asian snake wine – hard liquor in which a venomous snake, typically a cobra, and often also a scorpion, has been drowned – is similarly consumed to improve health and virility.

A POPULAR INGREDIENT

Centipedes, especially large-bodied species in the genus *Scolopendra*, are probably the number one animal ingredient in traditional medicines in the Far East. For example, dried centipedes are the most prescribed treatment in traditional Korean medicine, for anything from joint problems to stroke and snakebite. To meet the enormous demand for centipedes, South Korea imports large numbers of them. In 1991, for example, it imported more than 50 million centipedes. Interestingly, the Korean folk logic for prescribing centipedes to treat leg and foot problems is the remedy's leggy nature. The rationale of choosing a specific treatment on the basis of it resembling or being an antithesis of a patient's symptoms is widespread. For instance, Korean traditional medicine recommends the use of scorpions, which have a painful sting, for the treatment of pain, and blood-feeding horseflies for the treatment of blood problems.

Bundles of dried centipedes that are used in traditional Chinese medicine.

Other venomous arthropods used for therapeutic purposes in Korea include wasps, bees, ants, giant water bugs (Family Belostomatidae) and scorpions. The use of spiders in traditional medicine worldwide seems to be much rarer than other venomous arthropods. Traditional medicinal cultures not only use venomous animals, but also their venoms. Practitioners of Ayurvedic medicine, for instance, weren't just preoccupied with the treatment of snakebite, they were also pioneers in using snake venom as part of traditional medicine, mixing it with other ingredients to treat conditions as different as fever, coma, plague and tuberculosis. And two Amazonian tribes, the Ka'apor and the Sirionó, prescribe the stings of ants in the genus *Pseudomyrmex* for the treatment of fever and inflammation. Intriguingly, the venom of one species of *Pseudomyrmex* contains components that can suppress the activity of the human complement system, a part of our immune system that, among other things, can trigger inflammation.

The rationale used in traditional Korean medicine that tries to match a patient's symptoms to a proposed cure also underpins homeopathy. Preparations made with the venom of various species of snakes and spiders are routinely prescribed by homeopaths. For example, homeopathic medicine made with the haemorrhagic venom from New World pit vipers in the genus *Lachesis* (bushmasters) are prescribed for the treatment of blood-circulation problems, such as varicose veins. Venom from the timber rattlesnake, *Crotalus horridus*, is used to make homeopathic drugs used to stop haemorrhages, while medicines based on the neurotoxic venom of cobras (*Naja* spp.) or female black widow spiders, *Latrodectus mactans*, are recommended for treating angina-like symptoms. Of course, the successive dilutions used to prepare homeopathic drugs ensure that toxin molecules are scarcely, if at all, present in such preparations, but even if they were present at detectable levels, administering them via pills or drops guarantees that any active peptide toxins are denatured and deactivated in the patient's stomach acid. This is why venom-based drugs used in conventional medicine that consist of synthetic versions of peptides are never taken orally, but are always delivered directly into the bloodstream by injection or infusion.

For those who like a better chance that their venom-based medications contain at least detectable levels of ingredients, there are ointments and balms available for

Pseudomyrmex spinicola acacia ants. A unique class of peptide toxins, called myrmexins, have been found in the venoms of some *Pseudomyrmex* species. In humans myrmexins have pain relieving and anti-inflammatory effects.

the treatment of pain and inflammation made with cobra, viper or scorpion venom. People who would like to smooth out the wrinkles on their faces can buy creams, gels, oils, facial serums and face masks infused with the venoms of jellyfish, snakes and bees. Indeed, in a deliberate echo of the muscle paralyzing powers of Botox, some of the products with bee venom toxins are marketed as Beetox. Many snake venom cosmetics boast an ingredient called Syn®-Ake, a synthetic peptide of three residues that mimics a neurotoxin from the venom of the Southeast Asian temple pit viper, *Tropidolaemus wagleri*. It acts as an antagonist of the nicotinic acetylcholine receptor, preventing the reception of nerve impulses in muscles. This results in temporary muscle paralysis, which reduces wrinkles that are due to muscle contraction. But perhaps the most surprising, yet aptly named, cosmetic product that claims snake venom as an ingredient is marketed as Snake Oil for the treatment of hair loss.

Inducing altered mental states with venom

The use of venomous animals is not limited to treating ailments of the flesh or providing a balm for the vain. In various parts of the world they are used to induce altered states of mind for ritual or recreational purposes. In his book *Drugs in Afghanistan* David MacDonald describes how some Afghans smoke dried scorpions to produce highs that are more intense and longer than those achieved by smoking hashish. MacDonald interviewed a fisherman who, when in jail for illegal fishing, habitually smoked dried scorpions because they were the only source of psychoactive substance around. He also describes a heroin addict who allowed his pet scorpion to sting him to get high, but who also had to use painkillers simultaneously to make these experiences less unpleasant.

Afghanistan has also created a lucrative international black market for snake venom. In 2014, police in the European country of Moldova seized a shipment from Afghanistan with 9,000 vials of snake venom, worth an estimated 2.8 million euros on the drug market. The production of snake venom for the illegal trade is a particular problem in India, where smugglers have been caught carrying vials or condoms full of liquid or dried snake venom worth millions of dollars. However, those who are too poor to try milked snake venom might be able to get a fix via the more natural route of a snakebite. As Christie Wilcox recounts in her book *Venomous*, some Indian cities have snake dens where visitors can buy the risky opportunity to have venom delivered straight from fang to flesh. Wilcox also discusses the peculiar community of self-immunizers, who are venomous snake enthusiasts who self-administer dilute snake venom either to try and immunize themselves to the bites of their pets, or simply because it causes them to feel rejuvenated.

Direct envenomation is not just used to achieve cheap thrills. In some cultures the ritual administration of venom is the entry fee that boys and girls must pay to become adults. The use of venom in puberty rites is particularly well documented in several tribes of native peoples in the Americas. The chosen animals for most of these rites are various species of ants, whose stings can be intensely painful but are not generally dangerous. After their first menstruation, girls in the Amazonian Ka'apor tribe are subjected to an ant ordeal. After the girls' hair is shaven off, large termite raiding ants, *Neoponera commutata*, are strung along their forehead and torso, and allowed to sting them repeatedly. Justin Schmidt's 'tasting note' for the sting of this ant species, which scores a 2 on his pain scale (see chapter 3), is that it causes a debilitating migraine-like pain in the envenomated area. The Wayapí tribe, which lives 965 km (600 miles) from the Ka'apor, tie the same termite hunting ant species in strings to the arms, legs, abdomen, back and brow of girls that are to undergo the same initiation rite.

The venoms of ponerine ants, such as this *Neoponera apicalis*, contain peptides called ponericins, which have haemolytic, insecticidal, and antimicrobial activities.

The use of different species of ants in various types of initiation rites is widespread in Amazonia. Perhaps most famous is the use of bullet ants, *Paraponera clavata*, in virility rites in the central Amazonian Maués tribe (also known as the Sateré-Mawe tribe). Boys have to wear gloves loaded with bullet ants for minutes at a time, and withstand the excruciating pain of the most painful ant stings in the world. The resulting pain, hallucinations and fever can last for many hours, and the boys endure this ordeal many times on the long road to manhood.

Ant stings have also been used for competitive sport. One indigenous Californian tribe called the Northern Miwok held contests in which several men would lie down in a disturbed ant nest, and the last one to get up would receive an award from the chief. But ritual envenomations can be even more extreme than this. Anthropologist Kevin Groark has documented the widespread use of red ants as a ritual intoxicant by many ethnic groups of indigenous Californian Indians before the twentieth century. The red ants in question were California harvester ants, *Pogonomyrmex californicus*, with stings that are incredibly painful and long lasting. No insect venom is as toxic to vertebrates as that of harvester ants, and the California harvester ants are no different, with an LD_{50} value of 0.6 mg/kg. Like the undisputed leaders in vertebrate lethality, the Maricopa harvester ants, *P. maricopa*, we met in chapter 3, California harvester ant venom easily holds its own, gram for gram, with the lethal venoms of many dangerous snakes, including mambas, cobras, kraits and vipers. So what did the indigenous Californians do with these ants?

The native tribes of southern California considered venom to be a conduit to the supernatural. They believed that venomous animals, such as rattlesnakes, harvester ants or black widow spiders, could transfer shamanic powers via their bites and stings, and give access to so-called dream helpers in the spiritual realm. Everyone who hoped to live a long and healthy life required such a dream helper, and the way to find them was through vision quests. To induce hallucinogenic visions the indigenous Californians could choose one from their sacred trinity of powerful ritual medicines: tobacco, *toloache* or harvester ants. When tobacco is smoked or ingested in sufficient quantities it can assist in producing a trance-like state. *Toloache* is a plant, *Datura wrightii*, in the nightshade family that can induce vivid auditory, tactile and visual hallucinations when ingested. You may think that compared to smoking tobacco or ingesting *toloache*, getting high on ant stings is decidedly less attractive. You would be very right.

The ant ritual was so unpleasant that only men and boys who had gone through puberty were allowed to use harvester ants to visit their dream helper. The basic

Gram for gram, the venom of the 6 mm (¼ in) long California harvester ant, *Pogonomyrmex californicus*, is as lethal as that of many dangerous snakes.

procedure was as follows. For three days prior to meeting the ants, the spiritual traveller prepared by fasting and nightly vomiting. Then, under the guidance of a tribe elder, the traveller would lie down on his back, and be quasi force-fed large numbers of living harvester ants that were rolled in balls of moist eagle down to make them easier to swallow. Care had to be taken not to chew the ant balls. Then, after eating as many ant balls as they could – as many as 90, each with about five ants – the traveller was ready to embark on his spiritual journey. Their attending ant doctor would then startle, poke and agitate the travellers, in an attempt to rouse the ants to action. The ingested insects would bite and sting the traveller's innards until he would lose consciousness. This was the desired result. Having been stung unconscious, a successful vision quest would result in powerful hallucinations, in which the traveller would meet his dream helper in the form of an animal spirit, a personified force of Nature or a dead relative. The loss of consciousness was interpreted as a small death, in which the traveller was killed by the supernatural entities he sought to contact. Kevin Groark has calculated that the ingestion of 90 eagle down balls, with 4 or

5 harvester ants each, would correspond to 35% of the LD_{50} dose for a human of 45.5 kg (100 lb) if all the ants were to sting. A dose of venom such as this would surely be enough to cause mind-numbing pain and a loss of consciousness.

After regaining consciousness the spiritual traveller would drink hot water to induce vomiting to purge the ants from his system. He would then confer with his tribe elder about the meaning of his visions, and receive instructions about how to cement the relationship with the dream helper, for instance through prayer and ritual offerings of seeds, shell beads, tobacco or eagle down.

The vision quests of indigenous Californian Indians were designed to reach the magical world of spirits through the crude physicality of massive ant envenomations. These ant rituals were widespread and very similar even between ethnic groups with otherwise different cultures and different languages. Members of some tribes would repeat this excruciating procedure many times over several days. Men who sought to acquire shamanic powers would even undergo this ant ordeal many times over a period of months or years. Ritual ant ingestion was also used for therapeutic purposes, with the number of ants adjusted to the severity of the illness, and with the regurgitated ants serving a role in establishing the patient's prognosis.

Groark relates the harrowing experience of a Kitanemuk woman who suffered from a gynaecological problem after giving birth:

> 'They gave me red ants as a medicine, both externally and internally. First, they put them all over my belly, from navel down... so you could not see my skin for the ants. It is no trouble to get them to stay on, as they begin to bite and hang on very tightly the minute they touch your skin. The pain was intense. At the same time that the external treatment was applied, I also had to swallow a great many live red ants. I don't not know how many, a lot. They of their own accord cling together in balls and it is these balls of ants that you swallow. This was also very painful. They must surely have bitten me inside, as I felt like something was pricking me between my shoulders. Some days after taking the red ant treatment, a flow of blood came freely from my uterus, and I got well.'

Ants were even ingested as a form of preventative medicine to protect men from coming to harm, and members of one indigenous Californian tribe would ingest ants as a supernatural prophylactic to neutralize bad omens. It was also a widespread practice for adults to swallow ants to contact a dream helper to help cure a sick child.

The method of administering the venom, the dramatic physiological effects, and the spiritual goal, probably qualify the ant ordeals of the native Californians as the most extreme use of venomous animals on Earth. This and the homeopathic use of venom bracket a broad spectrum where venomous animals and their chemical cocktails are ingested, inhaled or applied externally, to heal, harm, beautify, get high, visit alternate realities or gain supernatural powers. Venom-based homeopathy and ritual mass envenomation also define opposite poles with respect to efficacy and possible placebo effects. However, all the uses of venomous animals discussed so far are relatively crude and indiscriminate. To truly exploit the power that lies buried in venoms, a more refined approach is required.

From venom to pharma

Many venom laboratories function as venom refineries. In the same way crude oil isn't good for much except killing seabirds, crude venom can hurt, maim and kill. But just like fractionating crude oil yields the useful products that power our cars and pave our roads, fractionating venom releases its greatest and most refined powers.

As we discussed in chapters 3 and 4, the power of a venom resides in the precise make-up of its toxin cocktail. Each toxin has its own individual characteristics that give it special affinities for specific molecular targets, and by binding to those targets it will exert specific physiological effects. In order to understand what causes the complex envenomation profiles that result from the simultaneous attack of multiple toxins, researchers unmix venom cocktails. Researchers interested in discovering and developing new drugs scan venoms and their toxin fractions to find effects that may be of medical interest. And as we will see, if they find something potentially promising, the game is on.

Venom prospectors are an optimistic breed. They have good reason to be. The venoms of the vast majority of known species of venomous animals haven't been investigated yet, and those that have, have already yielded powerful drugs for the treatment of a variety of serious conditions (see p.185). Some of these drugs achieve enormous annual sales. For instance, the Gila monster venom-derived diabetes drugs Byetta and Bydureon earned the pharmaceutical company AstraZeneca US\$635 million in revenue in the first three quarters of 2016. Yet, the road to such blockbuster drugs is long, expensive, and in no way guaranteed to succeed. It takes years and a raft of pre-clinical trials in animals followed by clinical trials in humans to develop a promising lead into a marketable new drug.

In the mid-1990s researchers discovered that a neurotoxic peptide isolated from the Caribbean sea anemone, *Stichodactyla helianthus*, potently blocks potassium ion channels. The presence of potassium channel blocking toxins in venom is valuable for predators as they help cause paralysis by hyperactiving nerves, which causes uncontrollable muscle contraction in prey. Such venom toxins have evolved in several groups of animals, including cnidarians, cone snails and scorpions. The sea anemone venom peptide was named ShK, which stands for _Stichodactyla helianthus_ K (the chemical symbol for potassium) channel toxin.

Two decades of research into ShK have yielded a synthetic version of the toxin, called Dalazatide, which is currently in development with KPI Therapeutics for the treatment of auto-immune diseases, such as multiple sclerosis and psoriasis. It has already successfully passed several clinical trials for the treatment of psoriasis in humans, and more studies on the treatment of other diseases are underway. These hugely promising results are the reward for many years of research, but how do researchers start on this trajectory? Often the first step is to find a promising envenomation syndrome. And for a venom prospector anything can look promising.

Physicians in Brazil are sometimes presented with victims of spider bites or scorpion stings that share one outstanding symptom: a painful and persisting erection known as priapism. The culprits are two of the country's most dangerous arthropods, the Brazilian wandering spider, _Phoneutria nigriventer_, and the Brazilian yellow scorpion, _Tityus serrulatus_. Envenomations by these animals are common. Sometimes they are fatal but sometimes they cause spontaneous erections, especially in boys and young men. This symptom peaked the interest of researchers because erectile dysfunction afflicts about a quarter of men worldwide under the age of 69. Erection-inducing venoms might just hold the secret to a future treatment for the 30–35% of men with erectile dysfunction who do not respond to Viagra or related drugs.

Envenomations by the Brazilian wandering spider and yellow scorpion cause a complex syndrome of symptoms, most of which are undesirable for any drug intended to improve people's sex lives, such as elevated heart rate, sweating, excessive salivation and vomiting. To get rid of these symptoms, researchers fractionated the venoms, and tested the refined toxins for specific effects. They discovered that the scorpion venom contains one, and the spider venom two neurotoxic peptides that, when injected into rodents, reliably cause priapism. Further research has revealed part of the likely mechanism by which these toxins act. One of the spider toxins, called Tx2-6, is currently best understood.

For something that most men achieve so easily, getting an erection is actually a fiendishly complex physiological process. In simple terms, to allow an erection, the smooth muscles in the erectile tissues and blood vessels of the penis need

Examples of drugs that are derived from venom proteins, which are in use or under development.

Generic name (brand name)	Species with the venom toxin	Medical condition treated	Mechanism of action	Biological role of original toxin
VENOM DERIVED DRUGS				
Captopril (Capoten)	Jararaca pit viper (*Bothrops jararaca*)	High blood pressure	Relaxes blood vessels	Aids prey capture by lowering blood pressure
Eptifibatide (Integrilin)	Pigmy rattlesnake (*Sistrurus miliarius barbouri*)	Acute blockage of blood flow to heart to prevent heart attacks	Prevents formation of blood clots (blood thinner)	Aids prey capture by facilitating internal bleeding
Tirofiban (Aggrastat)	Saw-scaled viper (*Echis carinatus*)	Acute blockage of blood flow to heart to prevent heart attacks	Prevents formation of blood clots (blood thinner)	Aids prey capture by facilitating internal bleeding
Batroxobin (Hemocoagulase, Defibrase)	Lancehead pit vipers (*Bothrops atrox* and *B. moojeni*)	Different types of bleeding and thrombosis	Activates blood clotting in the short term, but leads to break down of clots in the longer term	Aids prey capture by activating blood clotting
Exenatide (Byetta, Bydureon)	Gila monster (*Heloderma suspectum*)	Type 2 diabetes	Lowers blood sugar levels via various mechanisms, such as increased insulin secretion	Role in venom is unknown
Bivalirudin (Angiomax)	Medicinal leech (*Hirudo medicinalis*)	To prevent blood clotting during surgery	Inhibits blood clotting	Aids blood feeding by inhibiting blood clotting
Ziconotide (Prialt)	Cone snail (*Conus magus*)	Intractable, chronic pain	Blocks the transmission of pain signals	Causes paralysis of prey
VENOM DERIVED DRUGS IN CLINICAL TRAILS ON HUMANS				
ShK toxin	Caribbean sea anemone (*Stichodactyla helianthus*)	Autoimmune diseases such as psoriasis and multiple sclerosis	Inhibits inflammation-causing immune cells	Causes paralysis of prey
Chlorotoxin	Egyptian deathstalker scorpion (*Leiurus quinquestriatus*)	Used as marker for tumour cells	Binds to tumours of central nervous system	Causes paralysis of prey
Soricidin	Short-tailed shrew (*Blarina brevicauda*)	Various types of cancer	Inhibits cancer cell proliferation and kills cancer cells	Causes prey paralysis

to relax so that blood can flow into the penis. The chemical that triggers the relaxation of these muscles is nitric oxide, which is released by nerves in the penis. The neurotoxin Tx2-6 causes an erection by inhibiting the inactivation of specific sodium ion channels in these nerves, thereby prolonging the transmission of nerve impulses and the release of nitric oxide. When Tx2-6 is injected into mice or rats, it does exactly that. It reliably causes erections, even in elderly rodents, or those with high blood pressure or diabetes, to mimic the main risk factors for human erectile dysfunction. But there is a catch.

The Brazilian wandering spider, *Phoneutria nigriventer,* is an aggressive spider that is responsible for hundreds of bites in Brazil every year. Despite having potent neurotoxic venom, the vast majority of envenomations are mild.

The effects of venoms and toxins are dose-dependent. When mice are given Tx2-6, the first symptom to emerge is an erection. However, when the dose is increased, the animals start to develop some hair-raising symptoms. Their fur bristles, they hypersalivate and sweat, tears stream from their eyes, and they exhibit tremors and spastic paralysis, followed by death. When the mice are autopsied, it is found that the blood vessels of their major organs – kidneys, liver, lungs and heart – are severely congested. Death probably results from a stroke and bleeding in the lungs. Astonishingly, the rodential member emerges unscathed from this physiological wreckage. For optimistic venom prospectors, this is enough to work with.

In mice there is a narrow dose range that causes the desired effect without the unacceptable side effects. If the same were true for humans, it would be very risky to use a drug based on an unmodified version of Tx2-6. Humans are far less uniform in weight and physiological make-up than lab mice. It would be a dangerous challenge for a user of this drug to get the dose right without ending up in hospital. Scientists are therefore experimenting with making changes to the structure of Tx2-6 that may improve its potency and reduce unwanted side effects.

Sometimes a toxin can be tweaked to make it more suitable as a drug, but sometimes it can't. The painkilling drug Prialt is a synthetic version of a neurotoxic peptide from the cone snail, *Conus magus*, that is used to treat severe intractable pain. The pharmaceutical company that developed the drug spent two years tweaking and testing modified versions of the peptide to optimize its action and minimize its side effects, but to no avail. The version on the market today has the exact same residues as the snail's original toxin. The same is true for the anti-diabetes drug Byetta, which is a synthetic version of a Gila monster venom toxin. As it turns out, Tx2-6 can be tweaked to improve it for therapeutic use. In late 2015 scientists announced they had designed a synthetic peptide based on Tx2-6 that can cause rodent erections without the lethal side effects. This peptide even has pain killing effects, which makes it a promising candidate for further drug development.

Even when a venom-based drug is safe enough to be put through clinical trials, it may fail for lack of efficacy or the emergence of undesirable toxicity that wasn't apparent in pre-clinical trials on animals. For example, Exanta, a blood-thinning drug based on cobra venom and manufactured by AstraZeneca, was withdrawn from the market in countries where it had already been for sale after clinical trials to get it FDA approved in the USA found that it was toxic to the liver.

Chapter 7

A Microcosm of Venom

In this book we have introduced readers to the wondrous world of venom. The stories we have told about the diversity, power, evolution and human relevance of animal venoms offer a small sample of what is known, and represent an infinitesimal fragment of what science can ultimately discover. The trademark of these stories is diversity on every conceivable level. Venoms have evolved dozens of times, and underpin the life histories of countless species, each with its own specific structural and behavioural adaptations to effectively deliver the toxic cargo to prey, predators and competitors. Virtually all venoms are cocktails of different toxins that attack numerous physiological targets, causing multiple envenomation effects that range from subtle and reversible behavioural alterations to the catastrophic destruction of bodily integrity. And each unique blend of toxins has been recruited from multiple protein families to meet the specific challenges of its producer, as well as the common challenges faced by all venomous species.

This sheer diversity represents a key challenge for the development of the next generation of antivenoms that can be made in a laboratory without relying on the immune systems of herds of hoofed mammals. But it also defines the enormous

The venomous animal that humans have been in a most intimate mutualistic relationship with for thousands of years is the honeybee.

potential, some of it already fulfilled, of venoms for human exploitation in research, medicine, cosmetics and even agriculture. The vast majority of venoms haven't been studied, and many venomous species remain to be discovered in all the world's habitats. But even venomous species that are familiar to everyone, that have lived in close association with humans for millennia, and that have been the object of scores of scientific studies, still have many secrets to yield.

In this final chapter we will briefly retrace the general contours of the world of venom with a visit to a well-explored microcosm: honeybees. What we know about the unique venom system of honeybees captures in miniature our understanding of the biology, evolution and value of animal venoms in general.

Bee venom as a defensive weapon

Honeybees illustrate the ubiquity and proximity of the venomous world. Everyone has seen a honeybee. The Western honeybee, *Apis mellifera*, an insect native to Africa, the Middle East and Europe, was domesticated thousands of years ago, and we have since helped it colonize all continents, except Antarctica. Honeybees also exemplify the long and intimate relationship we have had with the world of venom, a relationship that in contrast to the one we have with such creatures as snakes and spiders, is voluntary and deliberate. Before we domesticated the honeybee, we harvested the resources locked in their nests in the wild, much as some traditional societies still do. Beekeeping has been documented at least since the ancient Egyptians 5,000 years ago, and it remains a widespread practice all over the world. This is not surprising. Honeybee nests provide a bounty of resources – beeswax, pollen, honey and nutritious eggs and larvae – that is unparalleled among social insects, attracting animal and human predators alike. Keeping such a large and valuable resource secure requires bee colonies to have formidable defensive capabilities. And they do. When the need calls for it, honeybee workers turn into suicidal assassins.

The honeybee venom system – its structure, the make-up of its toxin cocktail and the specific behaviours that deliver the venom – exemplifies the general principle that venom systems evolve from common building blocks. During the evolution of hymenopterans (the group including wasps, bees and ants), an ancestral ovipositor that was used both to lay eggs and inject venom, was modified into a specialized stinger that is used only to inject venom into prey or predators. The possession of such a specialized stinger is a diagnostic trait of bees, wasps and ants, and it

The Western honeybee, *Apis mellifera*, represents a microcosm of the world of venom, and reveals the general contours of the biology and evolution of venoms.

underpins the evolutionary success of a great radiation of more than 70,000 species of venomous stinging insects. The selection pressure most likely to be responsible for shaping the honeybee venom system is the need to defend the nest from large vertebrate predators, such as our primate cousins and us. Honeybees therefore need to be able to effectively inject their venom, causing an envenomation that is quick to take effect and sufficiently unpleasant to deter attackers.

When a large predator approaches a honeybee nest some guards immediately fly towards it, while others that remain on the nest extrude their stinger, raise their abdomen and flutter their wings. If this threat behaviour doesn't deter the enemy, the bees attack. The honeybee stinger is an ultra-sharp toxin-delivery structure that easily penetrates the skin of vertebrate predators. Pulling the stinger out requires two orders of magnitude more force than it takes to insert it because of its recurving barbs. Once a bee has stung and embedded its stinger into a victim, it easily rips out of the bee's abdomen, leaving behind an intact, venom-pumping defensive weapon. Although this kills the bee, the detached stinger delivers a maximum dose

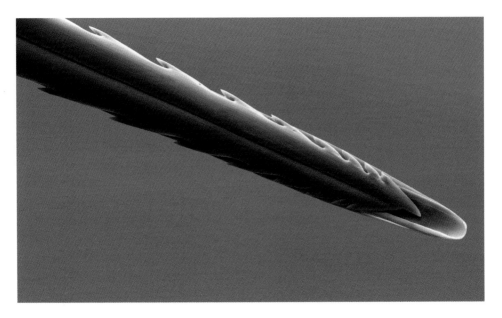

The tip of the stinger of a European honeybee is an exquisitely effective hypodermic venom delivery system.

of venom. Stinger-based alarm pheromones simultaneously recruit more suicidal sisters to the attack until the threat has been driven off.

Meanwhile, the dozen-and-a-half proteins that make up the honeybee's toxin cocktail attack multiple targets in the victim. The component that accounts for about 50% of the dry weight of honeybee venom is a unique peptide called melittin. It causes pain and tissue damage by activating pain receptors and by puncturing holes in cells, including red blood cells and nerves, thereby releasing pain-causing compounds. It is also a potent cardiotoxin that can damage the heart. The next two most abundant components of the venom are the enzymes phospholipase A_2 and hyaluronidase. These enzymes have been independently recruited into many different animal venoms and add to melittin's destructive powers. Phospholipase A_2 assists melittin in causing pain, but it also destroys cell membranes. On its own it is the most lethal toxin in honeybee venom. Hyaluronidase acts as a spreading factor that breaks down the extracellular matrix that maintains the integrity of tissues, which allows the venom to penetrate deeper into the victim. These two enzymes are also major allergens that can trigger severe or even lethal anaphylactic shock in humans.

Harnessing the power of bee venom

The exquisite constellation of structural, chemical and behavioural defensive traits possessed by the honeybee is also inspiring technological and scientific advances. Scientists are studying the unique mechanical properties of the honeybee stinger to try to develop microneedles for the painless delivery of drugs across the human skin. Apitherapy, the medicinal and cosmetic use of honeybee products, including their venom, has a history that spans millennia, from the ancient Egyptians to Gwyneth Paltrow. But treatments involving bee venom are not always without risk. The delivery of bee venom via acupuncture or the injection of dried venom frequently leads to adverse effects, including anaphylaxis. This is unsurprising considering what we know about this toxic cocktail, and it highlights the risks of administering crude animal venoms. Scientists are still deciphering the intricate physiological effects caused by honeybee venom both to better understand how to neutralize bee stings, and to investigate if bee venom can be used to treat medical conditions, kill pathogenic organisms, such as fungi and bacteria, or seed the development of new drugs.

For instance, despite the fact that bee venom phospholipase A_2 can cause anaphylaxis in humans, researchers continue to explore any beneficial effects it may have in animal models, including the suppression of inflammation. Another example is a small neurotoxic peptide in honeybee venom, called apamin, which is able to cross the blood-brain barrier and affect neurons in the central nervous system. Experiments in rodent models of Parkinson's disease, which is characterized by the progressive loss of neurons that produce the neurotransmitter dopamine, suggest that apamin may promote the survival of these neurons, which could slow the progression of the disease. But as with the neurotoxin from the Brazilian wandering spider discussed in the previous chapter, the therapeutic range of apamin is narrow. At higher doses it causes tremors and uncoordinated movements, which are precisely the symptoms it is hoped it would ameliorate.

This shows that, as always in the world of venom, any potential benefits are offset by risks. But although the health risks of bee venom injections are real, they shouldn't be exaggerated, at least not when a bee delivers them. In the USA roughly 100 people per year die of the stings of bees and related hymenopterans. In England and Wales on average only one person per year dies from a honeybee sting. Anaphylactic shock is the cause of death in virtually all of these cases, and this can be caused by one sting. In order to kill a person via non-allergenic effects, they would have to receive about 10 stings per pound of body weight (roughly 500 g or 17½ oz). This means that both authors should be able to survive at least 1,000 honeybee stings, but this is not an experiment we are willing to try.

The human benefits of bees are not limited to the lab. Like all venomous organisms honeybees occupy a specific niche in the economy of Nature that is in reciprocal dependence upon others that make up the ecosystem. To us bees are important because it just so happens that their niche overlaps substantially with our own. Foraging bees are crucial pollinators of many plants that we eat, and we therefore ferry bee colonies across the lands so they can pollinate the right plants at the right time. An American honeybee colony may start the year pollinating orange trees in Florida, then be shipped north to assist in squash pollination, and finish in Maine during the blueberry bloom, before starting another seasonal cycle.

It is only because our fates are so intimately entwined that we have become aware that honeybees are struggling. Chemical pesticides take their toll not only among the insects we don't want. They are diminishing honeybee populations at the same time as disrupting their normal behaviour. The world of venom offers some promising avenues to tackle this problem. Research done in recent years has focused on developing environmentally friendly bioinsecticides that harness the insecticidal powers of, for example, spider and scorpion venom toxins. Experiments show that insecticidal venom peptides can be fused to a carrier protein, which allows them to pass from the gut into the body of an insect that ingests them. This opens possibilities of developing crop sprays with such fusion proteins, or genetically engineered crop plants, or insect pathogens that can express such fusion proteins and thereby kill insects that eat them or become infected by them. Recent work done on a fusion protein containing an insecticidal toxin from the Australian funnel web spider, *Hadronyche versuta*, shows that it selectively kills a range of moth, beetle, bug and fly species, but it has no adverse effects on honeybees.

Chemical insecticides are not the honeybee's only problem. They are also under attack from a blood-sucking parasitic mite, *Varroa destructor*. Varroa mites can weaken and kill bees by drinking their body fluid and by transmitting pathogens. Varroa mites are native to Southeast Asia, where they are a natural parasite of the Eastern honeybee, *Apis cerana*. Assisted by our global transportation networks, the mites have spread across continents to decimate Western honeybee hives the world over. Varroa mites are currently the single biggest threat to Western honeybees in the world, a situation that is emblematic of our destructive impact on environments around the globe. Many other species will be equally impacted by our global activities, but the bulk of these go unnoticed. Honeybees are flag bearers for the overlooked multitudes of venomous animals that play central roles in local or global ecosystems, but that remain underappreciated, unstudied or unknown. In addition, they are the canary in the coalmine for monitoring negative impacts on the largely unseen natural foundation of human society.

The scientific frontier of the world of venom is long and convoluted, and located all along one side of this border lies the immense unrecognized value and future promise of venom. It lies hidden in every habitat, it is locked inside every venomous species, however well studied, and it stretches across all venomous branches of the tree of life. To illustrate some of the discoveries that venom research has yielded over the past few years let us return to honeybees one last time.

Varroa mites are venomous parasites. They pierce the cuticle of a bee or bee brood to drink their body fluid, and it was recently discovered that the mite's toxic saliva prevents proper wound healing so that it can feed on the bee's blood

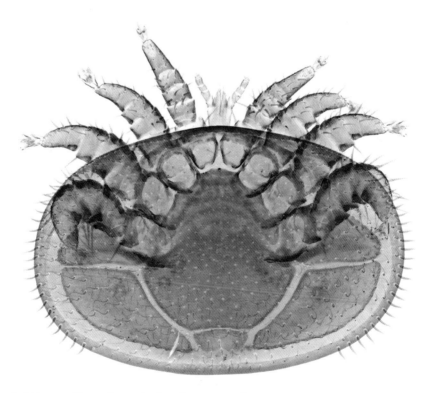

The arachnid bane of honeybee colonies: the varroa mite.

repeatedly. However, bees are not helpless against varroa mites. Honeybees groom each other and they can bite and remove varroa mites. Recent research shows that bees are able to deliver a lethal toxic bite to the mites. The bees' mandibular gland secretes a compound, called 2-heptanone, which paralyzes and kills varroa mites. Unfortunately, not all honeybee strains are equally vigilant when it comes to detecting and removing varroa mites from their hives, and many hives succumb to the blood-sucking parasite. Ironically, the solution may be yet another venomous animal. Arachnids called pseudoscorpions are naturally found in beehives and are predators of varroa mites. Most pseudoscorpions are venomous and can deliver venom to prey with their pincer-like pedipalps. Recent research suggests that these venomous arthropods show great promise as biological control agents of varroa mites when they are introduced into beehives. They attack and eat the varroa mites, but leave adult bees and their brood in peace. This trio of venomous arthropods –

A pseudoscorpion, *Chelifer cancroides*, on a fallen oak leaf. This tiny predator represents the unseen multitudes of venomous invertebrates that occupy all nooks and crannies of the natural world.

honeybees, varroa mites and pseudoscorpions – encapsulates the world of venom in a nutshell. They represent the good as well as the bad that the world of venom has to offer us.

No matter where you live, it is easy to underestimate how many times you cross the path of venom in a day. The jam you spread on your toast for breakfast exists by the grace of bees pollinating fruit trees. Ingredients derived from bee or snake venom may lace the moisturizing face mask you put on before going to work, and when eating a tasty calamari and fig lunch you are likely consuming the remains of two venomous animals (the female fig wasp often dies inside the fig after laying her eggs). When you drink a cup of tea in the afternoon you might enjoy realizing that your beverage and the sugar used to sweeten it, as well as the cotton clothes you are wearing while drinking it, are all made from plants protected from voracious insect pests by venomous parasitoid wasps that are deliberately used for this purpose. Then in the evening after dinner, if you need to take diabetes medication, you may well inject yourself with a synthetic version of a toxin from a venomous lizard, which may be interesting to ponder upon as you sip a glass of wine that has been clarified with the dried swim bladder of a venomous tropical catfish, *Arius maculatus*. These examples reveal only a fraction of the world of venom you willingly put on or in you on a daily basis.

The vast majority of venomous animals, such as pseudoscorpions, go about their business inconspicuously yet are all around us. They remain hugely underappreciated, perhaps in part because the world of venom is often presented in the media as just a collection of 'maxtremities', which exist solely for fearless TV presenters to fling themselves upon. Nature's ultimate weapon is really an ecological and evolutionary wonder that spans a truly mind-blowing diversity of species and more than half a billion years of history. And although most of us do not want to live with venomous animals, we certainly cannot live without them.

Glossary

alkaloid:
basic organic substances with one or more nitrogen atoms.

convergent evolution:
the independent evolution of similar organismal traits in different species.

defensins:
peptides found in plants, fungi, and animals that defend them against bacteria, viruses, and fungi.

disintegrins:
peptides that hinder blood clotting by inhibiting aggregation of platelets.

diversifying selection:
a type of natural selection that favours extreme values of traits in a population, leading to the divergence of these traits.

dopamine:
an organic chemical that functions as a neurotransmitter in the nervous system.

enzyme:
a molecule, most often a protein, that catalyses a chemical reaction.

ganglion:
a cluster of nerve cells or nerve cell bodies in the nervous system.

haemolymph:
the fluid circulating through the body of invertebrates such as arthropods, which carries out the functions that blood does in vertebrates.

hydrophilic:
the property of being attracted to water.

hydrophobic:
the property of being repelled by water.

ion channel:
a pore-forming protein that is inserted into a cell membrane and that regulates the flow of charged particles (ions) across the cell membrane.

neurotransmitter:
chemicals released by nerve cells that are necessary for the transmission of nerve impulses from one cell to another.

organic compound: chemicals with one or more carbon atoms linked to other atoms.

peptide:
a small protein often defined as consisting of less than 100 amino acids.

poison:
a toxic substance that causes dose-dependent physiological injury, and that reaches an organism passively, without the involvement of a special delivery mechanism.

potassium ion channel:
an ion channel that regulates the flow of potassium ions across cell membranes.

protein:
a molecule that consists of a string of amino acids.

protein domain:
a part of a protein that consists of a specific stretch of amino acids that folds into a specific three-dimensional structure and has a specific function.

purifying selection:
a type of natural selection that prevents deleterious changes in a trait from becoming fixed in a population.

scaffolds (evolutionary/structural):
stretches of amino acids that are crucial to maintaining a protein's 3-D structure; are evolutionarily conserved between species and function as a core around which evolutionary change can accumulate.

sodium ion channel:
an ion channel that regulates the flow of sodium ions across cell membranes.

toxin:
general descriptor for any substance that causes dose-dependent physiological injury.

venom:
toxic secretion produced by specialized cells in one organism that is delivered to another organism via a delivery mechanism – typically through the infliction of a wound – to disrupt normal physiological functioning to benefit the venom-producing organism.

Index

Further Reading

Books

Blackshall, S. 2011. *Venom. Poisonous creatures in the natural world*. New Holland, London. (Profiles of venomous and poisonous animals that occur on different continents.)

Fry, B. G. 2015. *Venom Doc*. Hachette, Australia. (Autobiography of a venom scientist who has gone to extreme measures to pursue the study of his lifelong passion.)

Fry, B. G. (ed.). 2015. *Venomous reptiles and their toxins. Evolution, pathophysiology and biodiscovery*. Oxford University Press, Oxford. (Comprehensive review of the composition, pathophysiology, and evolution of reptile venoms, as well as the methods used to study them.)

King, G. F. (ed.). 2015. *Venoms to drugs. Venom as a source for the development of human therapeutics*. The Royal Society of Chemistry, Cambridge. (State of the art review of the development of toxins into high value products such as therapeutics and biopesticides.)

Schmidt, J. O. 2016. *Sting of the wild*. Johns Hopkins University Press, Baltimore. (Unparalleled insight into the lives of stinging insects by the man who has devoted his career to their study.)

Wilcox, C. 2016. *Venomous. How Earth's deadliest creatures mastered biochemistry*. Farrar, Straus and Giroux, New York. (An intimate profile of venoms and their impact on humans.)

Journal papers and articles

Casewell, N. R. et al. 2013. Complex cocktails: the evolutionary novelty of venoms. *Trends in Ecology and Evolution* 28: 219–229.

Fry, B. G. et al. 2009. The toxicogenomic multiverse: convergent recruitment of proteins into animal venoms. *Annual Review of Genomics and Human Genetics* 10: 483–511.

Undheim, E. A. B. et al. 2016. Toxin structures as evolutionary tools: using conserved 3D folds to study the evolution of rapidly evolving peptides. *Bioessays* 38: 539–548.

Von Reumont, B. V. R. et al. 2014. *Quo Vadis* venomics? A roadmap to neglected venomous invertebrates. *Toxins* 6: 3488–3551.

Warrell, D. A. 2010. Snake bite. *The Lancet* 375: 77–88.

Picture Credits

Acknowledgements

We gratefully acknowledge the colleagues who gave us permission to use their images and photos in this book: Bjoern von Reumont, Volker Herzig, York Morgan, Nick Casewell, David Warrell, Alex Blanke, and George Madani. We thank Anna Smith for sourcing images, Celia Coyne for editing the text, and Trudy Brannan for efficiently steering the book through the editorial straits.

We thank Sarah Webb for reading the whole manuscript. RJ thanks Raph Chanay, Jess Simpson, and Bjoern von Reumont for their patience with his prioritization decisions. RJ gratefully acknowledges the U.K. Natural Environment Research Council (Grant NE/I001530/1) and the Biotechnology and Biological Sciences Research Council (Grant BB/K003488/1) for supporting his venom research.